IMAGES
of America

THE NAVY AT POINT MUGU

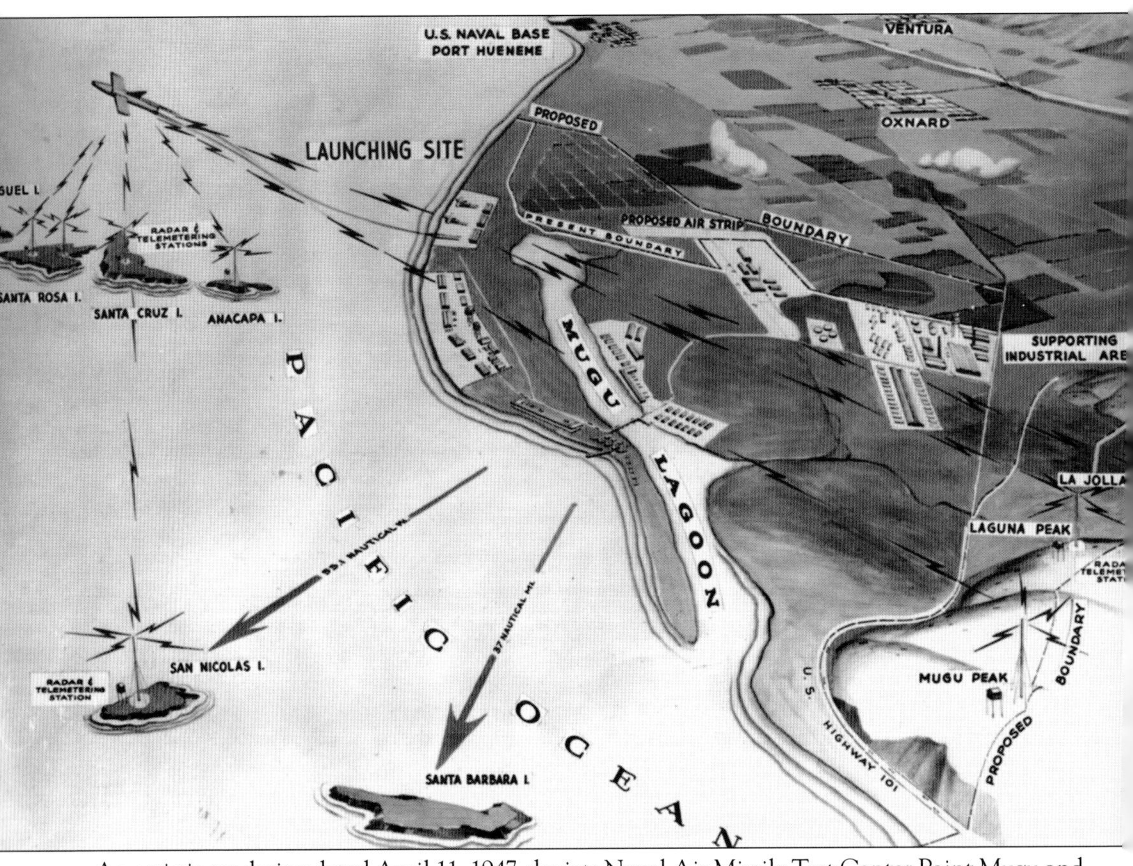

An artistic rendering dated April 11, 1947, depicts Naval Air Missile Test Center Point Mugu and its other areas of operation, including San Nicolas Island, Santa Cruz Island, Anacapa Island, and an operation post atop Laguna Peak. After recognizing the need to test Navy missiles over water, the Bureau of Aeronautics set up a small naval air facility at Point Mugu in 1945. (Courtesy of the US Navy Seabee Museum.)

On the Cover: A group of Air Development Squadron Four (VX-4) officers and enlisted men stand with their aircraft in the 1950s. Approximately 28 officers and 200 men formed VX-4, an operational squadron based at Point Mugu from 1952 to 1994. Besides conducting tests, evaluations, and investigating weapons systems, VX-4 developed the best ways to utilize an aircraft and its weapons for fleet operational use. (Courtesy of Naval Base Ventura County.)

IMAGES
of America

THE NAVY AT POINT MUGU

Gina Nichols

ISBN 978-0-7385-7532-2

Published by Arcadia Publishing
Charleston, South Carolina

Printed in the United States of America

Library of Congress Control Number: 2010942494

For all general information, please contact Arcadia Publishing:
Telephone 843-853-2070
Fax 843-853-0044
E-mail sales@arcadiapublishing.com
For customer service and orders:
Toll-Free 1-888-313-2665

Visit us on the Internet at www.arcadiapublishing.com

Contents

ACKNOWLEDGMENTS

The author wishes to thank her friends, family members, and son Hunter for their patience and support during this venture. Special thanks to Vance Vasquez, Naval Base Ventura County; the Point Mugu Alumni Association; and Bea Garcia, Command Archives, Naval Air Warfare Command Weapons Division Point Mugu for all their help in locating photographs and information essential to the writing of this book. Thanks to the US Navy Seabee Museum, the Naval Media Center's Navy Newsstand, the National Archives, and the Defense Visual Information Center for allowing me to scan photographs and obtain vital information about the history of Point Mugu from their collections.

The views, opinions, and statements presented in this book are those of the author and do not necessarily represent nor are they endorsed by the US government, the Department of Defense, or the US Navy.

INTRODUCTION

Point Mugu recorded history dates back to October 10, 1542, when Juan Rodriquez Cabrillo, the first European explorer to navigate the coast of present-day California, is thought to have landed at Mugu Lagoon as he charted the coastal waters. Cabrillo named the area Mugu after the native Chumash word *mu'wa* or *muwu*, meaning "seashore," "beach," or "hand of the beach."

The area surrounding Mugu was later devoted to agriculture, while Point Mugu became a popular recreation site. The area was used by a duck-hunting club when the sport of duck hunting came into fashion, and by the late 1920s, the Mugu Fish Camp was established for recreational deep-sea fishing. The fish camp featured a 350-foot-long pier and a string of small cabins for overnight use by local sportsmen.

In late 1942, the Advance Base Depot Port Hueneme needed space to train naval personnel in rapid landing operation involving LSTs and other landing craft. This specialized training required a beach of the same general type found on Pacific islands, and Point Mugu, located nine miles southeast of the base, provided the perfect auxiliary cam site. On February 9, 1943, the Acorn Training Detachment, later renamed the Acorn Assembly and Training Detachment (AATD), was commissioned to begin training Advance Base Air Units (ACORNs), Combat Aircraft Service Units (CASUs), and Group Pacific Units (GROPACs) to maintain and operate an advance air base. An ACORN was a small air base approximately one-quarter the size of a full air base, which was called an OAK. In March 1944, Seabees assigned to train at the Acorn Assembly and Training Detachment put down the section of Marston matting runway that became Point Mugu's first airstrip. The Seabees were established in 1942 to meet the Navy's wartime need for construction work in combat areas. The bases at Port Hueneme and Point Mugu, which were in dire need of buildings, utilities, and facilities, served as training opportunities for the Seabees before they were deployed to the Pacific.

During World War II, considerable progress was made in guided missiles, pilotless aircraft, and special weapons, leading to the need for bases specializing in the design, testing, and evaluation of new armaments. In late 1945, Naval Air Facility Point Mugu was established as a detachment of the Pilotless Aircraft Unit at Marine Corps Air Station Mojave. NAF Point Mugu gradually assumed greater importance with its program to test and evaluate the Loon surface-to-surface missile for potential use as a ship-to-shore bombardment weapon. Meanwhile, the Bureau of Aeronautics had recommended the Navy establish a missile test center. After surveying 59 proposed locations on the East, West, and Gulf Coasts, as well as in the Caribbean, a special committee recommended Point Mugu as the site for the new missile test center. On October 1, 1946, the Pilotless Aircraft Unit Mojave, Naval Air Station Mojave, and Naval Air Facility Point Mugu merged to become Naval Air Missile Test Center Point Mugu. The center also encompassed San Nicolas and Santa Cruz Islands for use with communications, weather reports, and documenting missile instrumentation.

In 1948, Congress earmarked $30 million for land acquisition and building construction for the new naval station and island facilities. On August 1, 1949, Naval Air Station Point Mugu was commissioned to provide infrastructure and personnel support to the Naval Air Missile Test Center (NAMTC). Since the mid-1940s, Point Mugu has had several center names, all with the mission to develop, test, and evaluate missiles and related systems. Missiles such as Oriole, Lark, Gorgon, Regulus, and many others have been developed and tested at Point Mugu.

Air Development Squadron Four (VX-4) moved to NAS Point Mugu in 1952 to conduct evaluations of air-launched guided missiles. Throughout its years at Point Mugu, the unit developed tactics and doctrines for the fleet, including the development of all-weather fighter/interceptor tactics for the use of air-launched guided missiles against all types of targets. VX-4 ended its tenure at Point Mugu on September 30, 1994.

Antarctic Development Squadron Six (VX-6) was established on January 17, 1955, to provide support for scientific exploration in Antarctica. VX-6 was redesignated VXE-6 on January 1, 1969, and, in 1974, the squadron's home base changed to NAS Point Mugu. During its service, VXE-6 transported more than 195,000 passengers, 240 million pounds of dry cargo, and almost 10 million gallons of fuel to sites in Antarctica.

In June 1958, the Pacific Missile Range, with Point Mugu as its central site, was established to relieve the Atlantic Missile Range of its heavy launching schedules at Cape Canaveral. The Pacific Missile Range was created to conduct research and development projects on the West Coast and test and evaluate programs. The range became the western member of the triad of national ranges, including the Atlantic and White Sands Missile Ranges.

In early 1958, the Army transferred more than 19,000 acres of South Camp Cooke to the Navy, which became Naval Missile Facility Point Arguello. The northerly 65,000 acres were converted into Vandenberg Air Force Base. Point Arguello became a component of the Pacific Missile Range and was used for missile launchings as well as NASA range operations. The base was involved in the launching and support operations of tactical weapons as well as programs such as Mercury, Thor, Atlas, Nike-Zeus, and Gemini. Point Arguello became Mercury Station No. 13 in 1961, and it monitored Mercury astronaut conditions and brought Enos—the first American primate launched into orbit—back from space on command. On April 9, 1963, the first operation in the Gemini series was supported by Gemini Station No. 13 at Point Arguello, which tracked it on the first four orbits.

In April 1962, the Navy Astronautics Group was formed at Point Mugu to carry on the development of *Transit*, a navigational satellite used by the Navy to provide guidance to navigators anywhere on or under the earth's oceans. The Navy Astronautics Group later managed a network of satellite tracking stations including determining orbital paths and injection of information for transmission by satellite to ships at sea around the globe. In June 1990, the Navy Astronautics Group was designated the Naval Satellite Operations Center (NAVSOC).

The Naval Missile Center (NMC) and Pacific Missile Range (PMR) merged on April 26, 1975, to become the Pacific Missile Test Center. As the defense budget decreased with the end of the Cold War, the Naval Air Systems Command positioned the bases at Point Mugu, China Lake, White Sands, and Albuquerque under the Naval Air Warfare Center Weapons Division. In January 1992, the Pacific Missile Test Center was disestablished, and the Naval Air Warfare Center Weapons Division Point Mugu was formed to streamline base functions under one command.

In 1998, four E-2C Hawkeye Squadrons transferred to Point Mugu from Marine Corps Air Station Miramar, California. The E-2C squadrons, along with the staff of Commander Airborne Early Warning Wing US Pacific Fleet, brought 16 E-2C Hawkeye aircraft and more than 1,000 personnel to Ventura County. Naval Air Maintenance Training Group Detachment, the final piece of the Hawkeye community to relocate from San Diego, moved to Point Mugu on October 17, 2000.

On October 11, 2000, the two naval bases of Naval Air Station Point Mugu and Construction Battalion Center Port Hueneme were consolidated into Naval Base Ventura County. Although the base name has often changed, the mission at Point Mugu has not faltered. The Navy continues to support many diverse military missions, including testing weapon systems, operating space satellite systems, and providing radar and communication support to the naval aviation community.

One

Camp Point Mugu

1943–1945

This May 5, 1943, view of the Point Mugu Fish Camp ticket office and bridge is from the main base towards the lagoon and beach. In 1943, the Navy sought an area that would provide beaches similar to those found on Pacific islands for the newly formed ACORN Training Detachment. This photograph was taken as part of the original land appraisal report. (Courtesy of the US Navy Seabee Museum.)

A pelican sits in front of former Mugu Fish Camp cabins at the ACORN Training Detachment on April 26, 1943. The cabins were transformed into officer housing for the duration of World War II. The Point Mugu camp at first consisted of no Navy-built construction, and Navy personnel were housed in these preexisting, privately built cabins and tents. (Courtesy of the US Navy Seabee Museum.)

Double and single cabins line the beach at the former Mugu Fish Camp on May 5, 1943. Rented for $2 or $3 per night, these lodgings lined the beach from east to west and were fronted by a wooden boardwalk. The Point Mugu Fish Camp became a popular spot for picnics, recreational saltwater fishing, and duck hunting. (Courtesy of the US Navy Seabee Museum.)

This view of the butane, fuel, and water tanks, a windmill, and a well on Juan Arias's ranch was taken on May 5, 1943. Initially, the Navy leased land belonging to the Mugu Fish Camp and Arias. Lots 2, 3, and 4, which were originally part of the Rancho Guadalasca land grant, were appraised by Tom Mason to estimate a value for personal property and improvements to the land. (Courtesy of the US Navy Seabee Museum.)

A May 5, 1943, interior view of the Point Mugu Fish Camp restaurant shows counters, stools, a coffee urn, and a cash register. The Navy initially intended to lease the Mugu Fish Camp and maintain the property in its initial condition. Improvements were made only if the structures interfered with naval operations. The landowners were compensated for their loss of property, land, and any improvements. (Courtesy of the US Navy Seabee Museum.)

Another May 5, 1943, view shows the deck and rails of the Point Mugu Pier. On September 28, 1939, the pier at the Point Mugu Fish Camp was destroyed and 26 people were swept to their death when the fishing boat *Spray* crashed ashore during a tropical storm. Construction of a new and longer pier, plus cabins and a restaurant, was completed in August 1940. (Courtesy of the US Navy Seabee Museum.)

This interior shot of the Point Mugu Fish Camp store shows the merchandise and countertop on May 5, 1943. The store, which also housed the owner's living quarters, sold all of the basic fishing, hunting, and convenience items necessary for its clientele: sportsmen. It was the only building to survive the 1939 tropical storm. (Courtesy of the US Navy Seabee Museum.)

An April 26, 1943, photograph shows Point Mugu Beach, Laguna Peak, and Dead Man's Rock at the newly established ACORN Training Detachment. An ACORN unit was an advance base naval air unit designed to quickly construct and operate a land or seaplane advance air base. Many of these units trained at Point Mugu before deploying to the Pacific theater during World War II. (Courtesy of the US Navy Seabee Museum.)

A 1944 aerial photograph of ACORN Assembly and Training Detachment facilities shows the pier, quarters, bridge, warehouses, and shops across the lagoon. Commissioned on February 6, 1943, ACORN Assembly and Training Detachment began operations with 12 officers and 56 enlisted men at Port Hueneme, California, and moved to Point Mugu in early 1943. Comdr. Marshall Gurney was the first commanding officer. (Courtesy of the National Archives.)

Built near what is now the Mugu Road Gate, a row of Quonset huts known as the "Dallas Hut area" was constructed in the summer of 1943 to house 1,500 men. By midsummer 1943, the remodeled private quarters along the beach were no longer adequate for the personnel stationed at Point Mugu, and construction of the first Quonset hut housing area began. (Courtesy of the US Navy Seabee Museum.)

This August 1, 1943, photograph is of Quonset huts at the ACORN Training Detachment area near Point Mugu, looking west from the road running into Point Mugu adjacent to Coast Highway 101. The marshy land at Point Mugu had to be drained and filled to ready the area for base personnel living quarters. (Courtesy of the US Navy Seabee Museum.)

Two unidentified Seabees survey the ACORN Training Detachment base for future expansion on August 1, 1943. Heavy equipment such as bulldozers, tractors, scrapers, dump trucks, and cranes were used in almost all phases of base construction and provided practical training opportunities for Seabees headed to advance base areas in the Pacific. (Courtesy of the US Navy Seabee Museum.)

In March 1944, Seabees from the ACORN Training Detachment laid the first permanent airstrip with Marston matting for the new airfield at Point Mugu. The original airstrip was 2,500 feet long but was later extended to 5,400 feet before the end of World War II. Marston matting is a standardized, perforated steel matting material originally developed for the rapid construction of temporary runways and landing strips. (Courtesy of the US Navy Seabee Museum.)

Members of ACORN 16 place two-foot-by-six-foot pontoons in front of an LST (Landing Ship, Tank) using an outhaul line and two bulldozers on August 8, 1943. In July 1944, construction of a "mock ship" school, a project for the training of personnel in cargo handling, was begun. Starting November 27, 1944, Navy personnel began cargo-handling training at the two mock LSTs built on the Mugu spit. (Courtesy of the US Navy Seabee Museum.)

As part of its advance base training on August 10, 1943, ACORN 16 attempts to reload a grounded LCPR (Landing Craft Personnel Ramped) by unloading outhaul anchors and emptying sandbags. The former Point Mugu Fish Camp pier is in the background. (Courtesy of the US Navy Seabee Museum.)

A row of Quonset huts is under construction at ACORN Training Detachment on August 17, 1943. Mess facilities, barracks, administration and recreational buildings, officers' quarters, an auditorium, classrooms, and drill fields were all being built. The area was called Splinter City due to its unfinished appearance. (Courtesy of the US Navy Seabee Museum.)

On September 1, 1943, Seabees attached to ACORN Training Detachment are constructing a water tank at Point Mugu. Plans for Splinter City began in April 1943, starting with the administration building, mess hall, and barracks. The last of the major structures to be planned was the main dispensary, which was designed and built in October 1943. (Courtesy of the US Navy Seabee Museum.)

A crane stuck on metal skids between two pontoons is pulled to shore by a bulldozer as part of advance base training on September 9, 1943. Seabees and ACORN personnel built berthing space for LSTs, small boats, and landing craft in early 1943. In June 1943, the LST training warehouses were built for the instruction of loading crews. (Courtesy of the US Navy Seabee Museum.)

ACORN personnel off-load aircraft from USS *Bismarck Sea* via a pontoon barge on July 24, 1944. In March 1944, the Secretary of the Navy had redesignated the name of the detachment to ACORN Assembly and Training Detachment to more accurately describe the complete function of the command that now included an airstrip for training personnel. (Courtesy of the National Archives.)

Naval personnel unload planes from the carrier USS *Bismarck Sea* on July 24, 1944. CASUs (Carrier Aircraft Service Units) used the Mugu airstrip for instruction in maintenance and upkeep of planes. Two hundred thirty-one carrier aircraft were unloaded directly from CVEs or brought in from airfields. The CASU unit was tasked with putting new life into discarded planes that once fought in the war or disposing them as scrap metal. (Courtesy of the National Archives.)

Navy aircraft returned from the Pacific war zone await repair at ACORN Assembly and Training Detachment Point Mugu in this image taken July 24, 1944. As part of their final training, an ACORN, CASU, and CBMU (Construction Battalion Maintenance Unit) were allowed to take over the airfield and operate it. This included refitting planes for use in the Pacific. (Courtesy of the National Archives.)

GROPAC 15 sets up a training camp at San Nicolas Island on May 22, 1945. San Nicolas and San Clemente Islands were used as training areas for ACORNs, CASUs, GROPACs, and Seabee units starting on July 7, 1944. A GROPAC (Group Pacific) unit supplemented an ACORN or other advance base unit by providing communications facilities, picket boats, anti-torpedo nets, harbor patrol facilities, and defenses. (Courtesy of the US Navy Seabee Museum.)

This c. 1945 image shows the hangar, parking area, and maintenance shops at AATD Mugu Airstrip built by the Seabees. The pierced plank, or Marston matting, airstrip was started as an airfield training project by the 101st Naval Construction Battalion and continued by other units. Adjacent to the airstrip, just over the small stream, a petroleum-oil-lubricant, or POL, tank farm was erected by the 30th Naval Construction Battalion. (Courtesy of the US Navy.)

Two

Birth of the Missile Age 1946–1957

Capt. Albert N. Perkins reads his orders and establishes the Naval Air Missile Test Center Point Mugu. The Pilotless Aircraft Unit (PAU) Mojave, Naval Air Station Mojave, and Naval Air Facility Point Mugu merged to become Naval Air Missile Test Center Point Mugu on October 1, 1946. Naval officers and enlisted men gathered to commission the new facility near the bridge over Mugu Lagoon. (Courtesy of Naval Base Ventura County.)

Base administration Building 1, designed to house most of Mugu's functional officers, is under construction in this March 25, 1949, photograph. In June 1948, Congress authorized $30 million to build a modern and complete missile-testing station at the newly designated naval base. Designed by the Parson's Aero-Jet Company and constructed by the A1-Co Company of Los Angeles, the building was slated to cost $355,000 and be completed by mid-1949. (Courtesy of the US Navy Seabee Museum.)

The massive dredge *Los Angeles*, the flagship of the Standard Dredging Corporation of Los Angeles, is shown conducting dredging operations in Mugu Lagoon on March 25, 1949. Before contracts could be advertised for new roads, utilities, and buildings, it was necessary to fill the low-lying area with three million cubic yards of material dredged from the shallow lagoon. (Courtesy of the US Navy Seabee Museum.)

This photograph is of the Naval Air Missile Test Center Point Mugu airstrip, taxiway, and hangars on August 24, 1948. Early construction at the airfield included additional aircraft facilities and renovation of the old hangars on the north side of the runways. (Courtesy of the US Navy Seabee Museum.)

VX-4 aircraft are seen on an aircraft parking apron on November 17, 1954. The primary mission of VX-4s involved the operational evaluation of air-launched guided missiles. Later renamed Air Test and Evaluation Squadron Four, it was one of oldest continuous tenant organizations until it was decommissioned on September 30, 1994. (Courtesy of the US Navy Seabee Museum.)

The original Marston matting airstrip is clearly visible in this mid-1940s photograph of a Douglas SBD Dauntless dive-bomber at Point Mugu. The SBD Dauntless was the standard shipborne dive-bomber of the US Navy from mid-1940 until November 1943, when the first operational Curtiss SB2C Helldivers arrived to replace it. (Courtesy of the US Navy Seabee Museum.)

The Project Loon launch area was located adjacent to the beach at Naval Air Facility Point Mugu, shown here on January 16, 1946. The Loon launch site was constructed as part of a Seabee training exercise in order to expedite construction. Seabees worked around the clock to complete the facility in seven weeks. (Courtesy of the National Archives.)

This 1945 photograph shows a Bat radar glide bomb, which was used by the US Navy during World War II. The Bat glide bomb, a predecessor to the Gargoyle, was put into operation against Japanese shipping during the late months of World War II. Termed "glombs," they were carried aloft and launched by a mother plane, then guided to the target. Gravity provided the propulsion. (Courtesy of the US Navy.)

This is the old military personnel housing area with new officers' quarters and barracks in the background in 1950. By the summer of 1943, the original sportsmen cabins that had been redesigned as officers' quarters were no longer adequate to house the officers. The Seabees constructed the first Quonset hut area with a capacity for 1,500 men. (Courtesy of the National Archives.)

The original Quonset hut barracks from the ACORN Assembly and Training Detachment are seen here on October 9, 1946. In late 1948, plans were drawn up for new personnel facilities, including barracks; a mess hall; a cold storage and bakery; quarters for married officers, enlisted men, and civilian technicians; and ship service and recreation facilities. (Courtesy of the US Navy Seabee Museum.)

An October 9, 1946, image shows the administration building and barracks at the former Amphibious Training Command area. The Amphibious Training Command administered and supervised the preliminary, amphibious, and special training of ACORNs and CASUs, as requested by Commander Air Force, Pacific Fleet. It also operated the Mugu airstrip with aviation personnel as an adjunct of ACORN and CASU training. (Courtesy of the US Navy Seabee Museum.)

An aerial view shows the Loon project area on October 9, 1946. Initially designed to be launched from the deck of a CVE aircraft carrier, the Loon was the Navy's version of the German V-1 buzz bomb. The first live firing took place on January 7, 1946, nine months before the Naval Air Missile Test Center was commissioned. (Courtesy of the US Navy Seabee Museum.)

Laguna Peak is pictured in 1945, shortly after the road had been built to the top and a site was prepared for early communications equipment. The location was brought into use in December 1950, when the Receiver and Transmitter Buildings were completed. Approximately 5.5 miles of new roads at the base and 3.5 miles of access road to the top of nearby Laguna Peak were completed by mid-1950. (Courtesy of the National Archives.)

Construction of enlisted men's barracks nears completion on August 27, 1952. By late 1952, two of the six barracks were finished with the other four in various degrees of completion. The barracks project included permanent technical buildings and installations as well as facilities for both San Nicolas and Santa Cruz Islands. (Courtesy of the National Archives.)

The concrete Flight Test Control Building looks austere while under construction on October 28, 1952. Designed by the Fluor Corporation, Inc., of Los Angeles, the structure was the nerve center of all missile flight operations. Congress appropriated an additional $6.1 million in 1952 for construction of facilities, including a launchpad, flight test control building, storage buildings, a power plant, and missile handling facilities. (Courtesy of the National Archives.)

Shown here in 1950 is the newly completed Administration Building that was constructed on the edge of the Dallas Hut area, one-half mile from the main gate. Originally, Building 1 was scheduled to house the commanding and chief of staff officers of the newly commissioned Naval Air Station Point Mugu, as well as the educational office, legal office, classrooms, conference rooms, and personnel office. (Courtesy of the National Archives.)

Naval Air Station Point Mugu's new runway, completed by the Griffith Company, is seen here shortly after the extension and resurfacing was completed. The runway, covered with asphaltic concrete, was extended to 5,500 feet long and 200 feet wide and could sustain aircraft up to 110,000 pounds. A new type of night lighting was installed at the same time. For two months during construction, the Operations Department moved to Oxnard Airport but continued flight operations. (Courtesy of the National Archives.)

The hangar and enlarged runways and taxiways are shown shortly after being completed in May 1950. Aviation facilities were a major item of construction under the initial congressional appropriations. The new 110,000-square-foot double hangar was completed in May 1950. The existing runway was enlarged to a length of 5,500 feet and a width of 200 feet, with a parallel taxiway with parking aprons. (Courtesy of Naval Base Ventura County.)

Seen here in 1949, Laboratory Evaluation Department Building 6-2 was located in "Six Area" at Point Mugu. Naval Missile Center laboratories produced information that helped the United States move men and machines farther into space. Equipment in the laboratories could subject biological specimens, missiles, spacecraft, and their components to wide varieties of conditions simulating numerous environmental extremes of launching and space travel. (Courtesy of Naval Base Ventura County.)

A c. 1950 view shows the Air Blast Facility combustion test tunnel. The facility was planned, developed, and constructed by center personnel. With an overall length of 135 feet, the tunnel assembly has four major sub-assemblies, an inlet bell, test-section nozzle box, and diffuser. The inlet bell is 20 feet across at the mouth and narrows to 63 inches as it joins the test section. (Courtesy of Naval Base Ventura County.)

A Sparrow I missile is set for launching at Naval Air Station Point Mugu around 1951. The Sparrow I was a short-range missile designed to ride a radar beam directed at the target by the launch aircraft. The first successful launch of a Sparrow I missile occurred on April 2, 1951, when the missile was fired from the Sparrow 35-degree, short-length ground launcher using a booster equipped with a JATO (jet-fueled assisted takeoff) rocket. (Courtesy of Naval Base Ventura County.)

The VX-6 aircraft crew arrives at Williams Airfield prior to takeoff on November 19, 1956. VX-6 deployed to Antarctica every year from October to February as part of Operation Deep Freeze and flew a squadron of LC-130 Hercules planes equipped to land on the ice and twin-engine UH-1N Huey transport helicopters, which brought supplies to the science stations. (Courtesy of the US Navy Seabee Museum.)

This c. 1950 view of the mess hall, and four of the six original barracks buildings in the last stages of construction, looks toward Laguna Peak and the Administration Building. The mess hall and six barracks buildings were constructed and designed by the A-1 Construction Company of Los Angeles as part of a $2.4-million contract, which was rapidly changing the appearance of the Naval Air Missile Test Center in the early 1950s. (Courtesy of Naval Base Ventura County.)

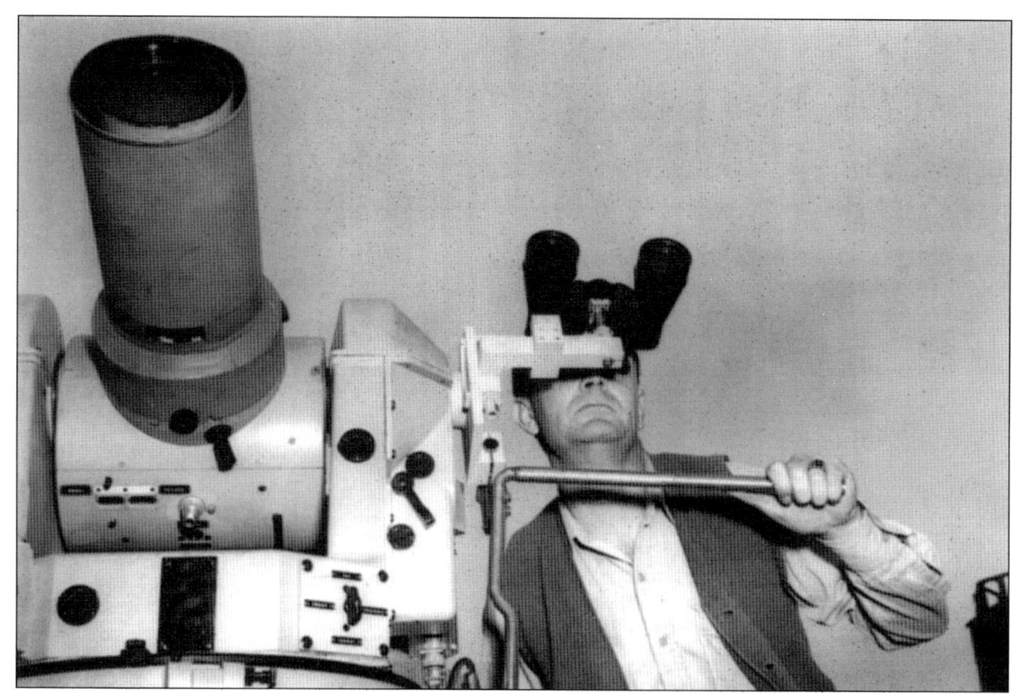

Seen in this 1952 image, an Askania Cinetheodolite station was located at San Nicolas Island. Six Theodolite stations were constructed that year, with four located on the shore at Point Mugu and two at San Nicolas Island. Askania Cinetheodolite is a camera-based system that measures and records azimuth, elevation, and time data on 35-millimeter film along with an image of a target. Multiple cinetheodolites operated together to track the target. (Courtesy of Naval Base Ventura County.)

A radar van is situated at the Camarillo airstrip around 1951. In 1951, the Naval Air Missile Test Center created the first frequency interference and control section. The entire department consisted of three people, part of the receiving building, an old panel truck, and shared use of an area-clearance aircraft. Eventually, World War II–era radar trucks were adapted for program use. (Courtesy of Naval Base Ventura County.)

In this October, 9, 1946, photograph, the ACORN Assembly and Training Detachment Beach Camp Area are located in the old fish camp cabins. The headquarters for ACORN Assembly and Training Detachment resided in the long structure in the foreground. The area later housed the Technical Service Shops, the Photograph Lab, and segments of the Component Test, Industrial Relations, and Technical Service departments. (Courtesy of the US Navy Seabee Museum.)

The original ACORN Assembly and Training Detachment airstrip, hangar, and facilities used in World War II are seen on October 16, 1946. In 1944, the ACORN Training Detachment had requested advance-base components to run the field as an actual overseas base. It was a subsidiary of NAS San Diego since all Navy airfields had to be under the Naval Air Center. (Courtesy of the US Navy Seabee Museum.)

A 1948 view shows Naval Auxiliary Air Station San Nicolas Airstrip, known as "the Rock." In 1950, NAAS San Nicolas operated on a 24-hour schedule in maintaining tracking service for guided missiles. Manned by a handful of naval personnel and civilians, the facility's activities included running radio and weather stations, maintaining the airstrip, and operating emergency rescue crews. (Courtesy of the US Navy Seabee Museum.)

The theodolite station built on San Nicolas Island is pictured in its uncovered position on August 11, 1948. The theodolite had a long track to move it far enough away from the building to provide an unobstructed view. After construction completed on the island, San Nicolas was used for the first time for communications, weather reports, and instrumentation. (Courtesy of the US Navy Seabee Museum.)

A view looking west shows upstream test air piping and the south side of the wind tunnel building shortly after construction was completed on May 1, 1951. The wind tunnel operated by a mixing principle. The primary air from the supercharger burners passed through the annular nozzle and mixed with the air at the diffuser inlet causing suction through the test section. (Courtesy of the US Navy Seabee Museum.)

A view looking northwest shows the Small Missile Project Building shortly after it was completed on December 8, 1950. The structure was a two-story, steel-frame building used to test, evaluate, and research weapons systems. The second floor provided office space while the first was used for laboratory and workspace. (Courtesy of the US Navy Seabee Museum.)

The new transmitter building (right, foreground) and parking area for mobile equipment (top, center) are located at Laguna Peak, seen here on July 16, 1951. The peak, approximately 1,500 feet above sea level, was brought into use by the center in December 1950, when the Receiver and Transmitter Buildings on the summit were completed. The structures were manned by 11 civilian and military personnel of the Field Instrumentation Division and Flight Test Control Division. (Courtesy of the US Navy Seabee Museum.)

Seen here on December 8, 1950, the Test and Evaluation Building (foreground), Small Missiles Project Building 1 (left background), and the hangar (right background) were built under the early construction program to create a missile test center. The base laboratories produced simulated environmental conditions including shock, vibration, and most weather conditions. (Courtesy of the US Navy Seabee Museum.)

A Regulus I surface-to-surface guided (cruise) missile is seen after launch from a Point Mugu launchpad around 1953. Regulus I came to the Naval Air Missile Test Center in 1947 as a model drop test program. By 1953, the Regulus program had accumulated enough flights to begin early fleet testing of the missile with fighter aircraft. (Courtesy of the National Archives.)

A Loon missile is prepared for launching on the AT-1 powder catapult at Point Mugu Beach around 1945. The first live firing took place on January 7, 1946, but the pulse jet engines failed during acceleration, and the missile glided for a mile before splashing into the ocean. After two years of testing, research, and development, the Loon was deemed ready for launch testing from a naval vessel. (Courtesy of the National Archives.)

One of the first buildings finished at the new Naval Air Missile Test Center around 1950 was a standard 110,000-square-foot double hangar for the aircrews. Miscellaneous airfield buildings were also constructed, including aircrew ready rooms, ammunition storage, a paint storeroom, and pyrotechnic storage. (Courtesy of Naval Base Ventura County.)

This c. 1949 image show the Lark radar guided missile during preparation for launching. The Lark antiaircraft missile program began in late 1944 to advance the effort to fight Japanese kamikaze pilots. Testing the Lark from shipborne launchers began in 1950 on the missile test ship USS *Norton Sound*. The Lark antiaircraft missile program was terminated in late 1950. (Courtesy of Naval Base Ventura County.)

This January 28, 1952, photograph looks eastward, showing the Air Blast Facility and the pier supporting a 36-inch saltwater intake line (right) and Laguna Peak (background). The Air Blast Facility could simulate conditions at speeds of nearly 600 miles per hour at altitudes to 10,000 feet. It was designed and constructed by center personnel and was used for testing air engines and other parts of missiles at subsonic speeds. (Courtesy of the US Navy Seabee Museum.)

A view looking northwest shows the main base Transmitter Building and the 150-foot steel tower on July 16, 1951. Because of propagation fading and unreliable equipment, the center upgraded its communications equipment by installing a new transmitter and receiver buildings at Point Mugu, Laguna Peak, and San Nicolas Island. (Courtesy of the US Navy Seabee Museum.)

A May 29, 1953, view looking northeast shows the High Pressure Air and Inert Gas Station in the foreground and the Test and Evaluation Building beyond. Both of these structures were used early in the Test and Evaluation Program to try out missile components at varying extremes of altitude, temperature, speed, pressure, vibration, and shock. (Courtesy of the US Navy Seabee Museum.)

San Nicolas Island is seen in 1947, just prior to the runway being extended to 8,000 feet in order to accommodate heavier and faster types of aircraft. Additional construction included a breakwater and pier. The islands were equipped with precise radar, optical, and photographic tracking devices, ensuring a continuous and accurate record of missile courses. (Courtesy of Naval Base Ventura County.)

The Point Mugu area is shown on August 1, 1948, prior to the Navy beginning construction on its huge permanent guided missile test center. Point Mugu was chosen over 26 other possible sites because the Santa Barbara Channel Islands afforded excellent bases for radar and tracking devices to study the flight of missiles as they hurled towards theoretical targets at sea. (Courtesy of Naval Base Ventura County.)

The Naval Observation Site is under construction on Santa Cruz Island on August 24, 1949. The Navy began using the island that year, when it leased a 10-acre tract and established communications, optical tracking systems, and radar surveillance facilities. Over the years, technology was updated and, by the mid-1960s, a telephone exchange center to the mainland was installed. (Courtesy of the US Navy Seabee Museum.)

This July 13, 1954, photograph shows the Supply Department Building 65, with the Public Works Office and shops in the background surrounded by empty lots. At the time of construction, the Public Works Department performed engineering and design services; maintained 800 structures, utilities systems, and roads; and provided transportation vehicles covering everything from on-base taxi service to heavy-duty trucks. (Courtesy of the US Navy Seabee Museum.)

Lark launchpad Baker is shown under construction at Point Mugu in December 1947. Rocket powered, the Lark was designed to streak from the deck of a ship or from a ground installation and intercept an attacking plane formation. Beginning in 1950, the Lark was tested off the USS *Norton Sound* near Point Mugu, where the first interception of a moving air target by a surface-to-air missile was achieved. (Courtesy of Naval Base Ventura County.)

The original ACORN Assembly and Training Detachment personnel area near the mouth of Mugu Lagoon with the Air Blast Facility in the right center is seen on September 28, 1949. The lagoon area contains three basic sections, which are ecologically distinct yet interdependent. (Courtesy of the US Navy Seabee Museum.)

On July 29, 1949, the center is seen shortly after new construction began, and the fill area was completed. Congress appropriated $6.1 million in 1951 and another $9.9 million in May 1952 for further construction at the center, including erecting six more barracks, an additional test and evaluation building, a civilian cafeteria, bachelor officer quarters, a mess hall, supply building, and other base facilities. The Bureau of Aeronautics later requested an additional $7.8 million. (Courtesy of the US Navy Seabee Museum.)

The main tidal and cultivated lands located at Point Mugu are seen with Laguna Peak in the background in 1948. Because a large portion of the base was barely above sea level, one of the first tasks was to fill the low-lying areas. Mugu Lagoon was dredged, and the sand was transported via a large pipe to low areas. Two dredges worked simultaneously to expedite the project. (Courtesy of the US Navy Seabee Museum.)

On June 18, 1953, the first attempt is made to fire the Regulus missile from a short-rail launcher on the submarine USS *Tunny*. Later, a portable unit was placed on aircraft carriers but failed. Next, launching the Regulus from a cradle with an aircraft steam catapult was tested on the newly refitted USS *Hancock*. (Courtesy of the National Archives.)

The first Loon missile test launches from the USS *Cusk* ramp launcher using rocket boosters on December 2, 1946. The Bureau of Aeronautics believed in the Loon program despite its low success rate and, by the end of 1946, extended the project to include testing the Loon from a surfaced submarine. The USS *Cusk* was chosen for the test vessel and a 90-foot rocket-launching ramp was design, constructed, and installed on it. (Courtesy of Naval Base Ventura County.)

This view of the Target Squadron hangars and Point Mugu airstrip is from 1953. Nearly as important as the missiles themselves are the targets to test the weapons' accuracy. In the case of air-to-air missiles, target drones represent actual aircraft the missile is designed to attack; they mimic the speed and the reflective surface of the target. (Courtesy of Naval Base Ventura County.)

A Loon missile is mounted on the McKierman Terry XM-1 catapult launcher at Point Mugu around 1948. The Loon launch site was designated a Seabee homeport training exercise in order to expedite construction. The first McKierman Terry XM-1 launcher arrived in August 1945 and was initially used to test dead load shots. By January 7, 1946, the first live firing took place. (Courtesy of the National Archives.)

The first successfully launched Loon missile roars into the air with the assistance of four Monsanto rockets on July 15, 1947. The Navy showed off its guided missile Loon on the beach at Point Mugu to a crowd of VIPs. The rockets and sled fell free at the expiration of the thunderous thrust, which lasted two seconds. (Courtesy of the National Archives.)

Seen here in April 1952, test pits E and F were part of the rocket test run used by center personnel to try out individual missile parts before they were incorporated into the weapon. It was realized early on that, because of their vital interdependence, all components of a guided missile needed to function perfectly for a successful flight, and pretesting was an essential part of missile production. (Courtesy of Naval Base Ventura County.)

This c. 1950 view shows the main fire station and the old Dallas Hut housing area. By midsummer 1943, construction began on the first Quonset hut area, with a capacity of 1,500 men. After the war, the Dallas Hut area transitioned into Homoja housing for personnel and their dependents; this enabled two two-bedroom units in each 20-foot-by-48-foot Quonset hut. (Courtesy of Naval Base Ventura County.)

Seen here around 1950, the original operation, test, and evaluation area was at the Point Mugu spit. The Test and Evaluation Department formed in 1947 to test missiles and simulate air launching to gauge hazards in newly developed missiles. Laboratory techniques proved individual parts before they were incorporated into the missile and tested completed weapon systems. (Courtesy of Naval Base Ventura County.)

A 1951 view shows the early base layout of roads and buildings just starting to take shape. The image includes the new asphalt airstrip with new and old operations area on either side, the new hangar under construction, the small missiles building, the test and evaluation building, and the range instrumentation building. (Courtesy of the US Navy.)

Members of VX-4 pose with their aircraft in the mid-1950s. VX-4 conducted tests, evaluations, and investigations of aircraft weapons systems, support system equipment, and materials in an operational environment. After all the laboratory evaluations were completed, the final proof required of a guided missile was flight evaluation. For this test, the air-launched missile was cradled beneath its aircraft, carried aloft, and fired. (Courtesy of the National Archives.)

Seen here in March 1956, the launching area was the largest and most costly of the early projects. The testing process transformed from the early "launch and look" method to exhaustive tryouts of individual components, followed by laboratory studies, ground firings, and captive flights. The permanent launching area facility consisted of missile launching pads, flight test control, and missile support buildings, roads, and utilities. (Courtesy of the US Navy.)

The interim Induction Air Blast Facility at the center could simulate speeds of near 600 miles per hour at altitudes to 10,000 feet in the testing section, shown shortly after completion in 1950. The facility, designed and constructed by center personnel, was used for testing engines and other parts of missiles at supersonic speeds. In the background and still under construction is the Air Blast Facility, which provided supersonic tests. (Courtesy of the US Navy.)

A Loon guided missile, powered by a screaming pulse jet engine, is seen just before it is to be launched from the track launcher around 1948. Developed from Germany's deadly buzz bomb, the stub-winged Loon incorporated radio control, which was designed to give it exceptional accuracy at the 100-mile range. However, this missile hurled 100 yards off the launching track and dove into the surf. (Courtesy of the National Archives.)

The first Regulus II missile begins its flight down the Pacific Missile Range from the Building 55 launchpad in 1958. The Regulus II was a ship- and submarine-launched, nuclear-armed cruise missile deployed by the Navy from 1955 to 1964. At 57 feet long and weighing 21,000 pounds, the Regulus II achieved boost at launch with a rocket motor and a 10,000-pound thrust turbojet engine to reach a speed of Mach 2. (Courtesy of the National Archives.)

An aerial view shows Naval Air Station Point Mugu, originally a component of the Pacific Missile Range, which maintained and operated the support facilities at Point Mugu and San Nicolas and Santa Cruz Islands. Above and beyond the normal scope, Naval Air Station Point Mugu also provided operations, surface craft, and public works and operated the outlying field at San Nicolas Island. (Courtesy of the National Archives.)

Three

The Space Age Arrives at Point Mugu 1958–1969

The new Los Posas Gate is under construction on June 7, 1960. The new entry, adjacent to the Los Posas Road and what was then California Highway 101A, was opened after the California Highway Department completed the extension overpass providing a new exit and entry for heavy morning and afternoon traffic between Point Mugu and Camarillo. (Courtesy of the US Navy Seabee Museum.)

An image from February 11, 1960, shows the Component Test Building at Naval Air Station Point Mugu, where the parts of a system were brought together for the first time, assembled, and then underwent testing by computer and environmental simulation. Once parts passed the test, they were assembled to form a completed missile and were then retested. (Courtesy of the US Navy Seabee Museum.)

The aircraft parking apron and maintenance hangar, used for drone aircraft, is seen on January 17, 1962, while under construction during the second large wave of base renovation. In the case of air-to-air missiles, the target is a drone, an unmanned aircraft flown by remote control from ground stations or other aircraft. (Courtesy of the US Navy Seabee Museum.)

The battery of enlisted men's barracks is lined up in succession at Naval Air Station Point Mugu on February 25, 1959. The naval air station was responsible for one of the biggest housekeeping jobs of any activity in the Navy. It had to house, clothe, and feed more than 3,000 personnel assigned to the base and maintained nearly 1,000 structures at Point Mugu and offshore island establishments. (Courtesy of the US Navy Seabee Museum.)

The enlisted personnel barracks and new bachelor officer quarters (BOQ) are under construction at San Nicolas Island on June 30, 1959. As the range expanded its capabilities, new equipment and facilities were added and improvements were made to the original structures at all the island facilities. (Courtesy of the US Navy Seabee Museum.)

The Center Theater, seen here around 1958, was one of the most important recreation activities on the base, especially in the early years, and was where movies, cartoons, and short subject films were shown. The films shown at the theater were obtained from the Navy Motion Picture Exchange in Los Angeles, which handled all the film distribution for military installations in Southern California. (Courtesy of the US Navy.)

Seen here on March 15, 1969, Building 375, the headquarters for the Navy Astronautics Group, was just one part of its operations located at Point Mugu, which include a space tracking facility, a large electronic digital computer center, and a communication facility at Laguna Peak. The Astronautics Group was created to manage the Navy's space systems, including the Navy Navigational Satellite System. (Courtesy of the National Archives.)

Taken on March 15, 1969, this photograph shows Building 36, the headquarters for the Pacific Missile Range, which was created to conduct research and development projects on the West Coast and to test and evaluate programs. With the advent of more complex missile systems and the introduction of more guided missile ships and aircraft to the fleet, the problem of firing ranges became more important to the Navy. (Courtesy of the National Archives.)

This March 13, 1961, view shows the airstrip after the main runway extension was completed. Lengthening the main runway to 11,000 feet afforded adequate room for practically every type of aircraft in landings and takeoffs. Air Force One used the Point Mugu airfield to fly in US presidents John F. Kennedy, Richard Nixon, Ronald Reagan, and George H.W. Bush to visit military forces and the Southern California area. (Courtesy of Naval Base Ventura County.)

Tuffy the porpoise is ready for transport in 1965. The Navy Marine Mammal Program (NMMP) at Point Mugu studied the military use of marine mammals and trained them to perform tasks such as ship and harbor protection, mine detection and clearance, and equipment recovery. The animals' intelligence, exceptional diving ability, and trainability led to the founding the program in 1962. (Courtesy of the National Archives.)

Fronting Point Mugu Lagoon, the Marine Biology Facility, seen on October 5, 1964, used the mile-long body of water in its work with porpoises and other marine animals. It served as a transitional training area between research tanks and the open sea. At left foreground is the MacGinitie Marine Aquarium and Lab, with specimens of sea life from the lagoon and surrounding waters. (Courtesy of the National Archives.)

Porpoise 1st Class Tuffy is being removed from a small research water tank in 1965. A major accomplishment of the Navy Marine Mammal Program was the discovery that trained dolphins and sea lions could reliably work untethered in the open sea. In 1965, Tuffy participated in the Sealab II project carrying tools and messages between the surface and the habitat 200 feet below. (Courtesy of the National Archives.)

Sam the sea lion and his trainer, Wally Ross, head down the ramp at the Marine Biology Facility for a workout in the lagoon on March 16, 1965. Sam was being taught to recover an object placed at increasing depths in the open sea. The training was also used to study the diving physiology of sea lions. (Courtesy of the National Archives.)

A Point Mugu maintenance hangar is under construction on January 6, 1960. The maintenance facility has five major functional areas: the main hangar, parts room, engine shop, avionics shop, and main office. In addition, the machining shop and a paint shop were also available to the technicians. (Courtesy of the US Navy Seabee Museum.)

The Mercury tracking station is seen here on April 23, 1963. In 1960, the tracking station was built on Canton Island, now part of Kiribati, and the facility was utilized through November 1965 as part of the Mercury program. The US Air Force and the US Space and Missile Systems Organization continued to use the site for missile-tracking operations through 1976. (Courtesy of the National Archives.)

New maintenance hangars, called Buildings 353 and 365, are seen on November 27, 1961, shortly after construction was completed. The Aircraft Maintenance Department was responsible for class C maintenance of all jets and reciprocating aircraft assigned to the Naval Missile Center, the Pacific Missile Range, and the Naval Air Station Point Mugu. (Courtesy of the US Navy.)

In 1962, Pacific Missile Range instrumentation vans, buildings, and radar were located at the Tern Island Facility, French Frigate Shoals, in the Hawaiian Islands. The use of telemetry in gathering data during missile launches required global sites including those at Tern Island, San Nicolas Island, Santa Rosa Island, Point Arguello, Kokee Park, Barking Sands, South Point, Kwajalein Atoll, and Canton Island. (Courtesy of the National Archives.)

The Atlas missile launch site, which was a self-propelled launching gantry at Naval Missile Facility Point Arguello—a major launching site of the Pacific Missile Range—stretches 135 feet into the sky on March 1, 1960. The gantry is still in use today as part of Vandenberg Air Force Base. The base was the largest piece of real estate in the Pacific Missile Range complex and was the range's primary launch site for many years. (Courtesy of the US Navy Seabee Museum.)

The area surveillance room at Naval Missile Facility Point Arguello, seen below on April 28, 1961, was one of the many facilities operated by the Pacific Missile Range, which supported missile launches from the base. From this room, ships, aircraft, personnel, and trains were charted, and the area around the launch path was cleared prior to liftoff. (Courtesy of the National Archives.)

Point Arguello's gantry towers are shown on the base launchpad around 1962. The base was a perfect location to fire missiles and rockets because the crevassed terrain provided natural barriers for operations, and it was isolated from the nearest town by a range of mountains. A total of $100 million was invested into the 20,000 acres that were part of NMF Point Arguello. (Courtesy of the US Navy Seabee Museum.)

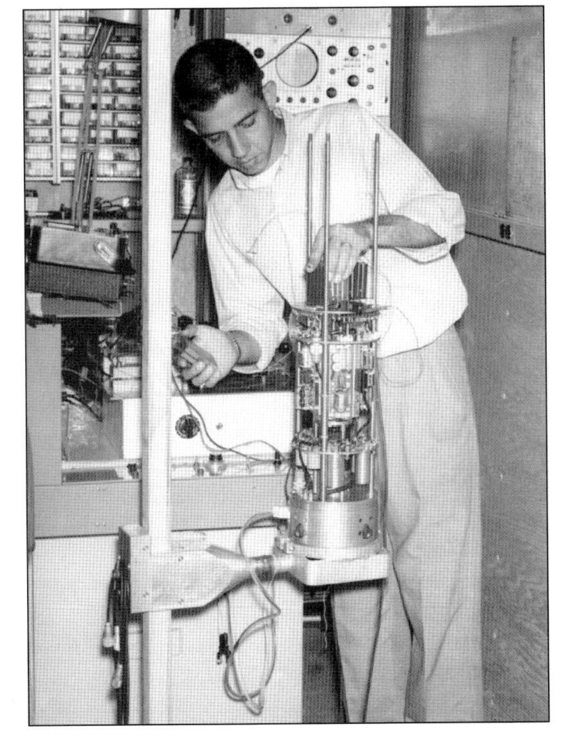

On September 20, 1960, an unidentified member of the Sunflare Project team adjusts the transmitter to be sent aloft by the naval research lab to record radiation data. Between July 14 and August 31, 1959, five Nike-Asps were launched as the first part of the solar x-ray mission by the US Navy. In an unsuccessful effort to measure lunar x-ray emission, the last Nike-Asp launched on September 27, 1960. (Courtesy of the National Archives.)

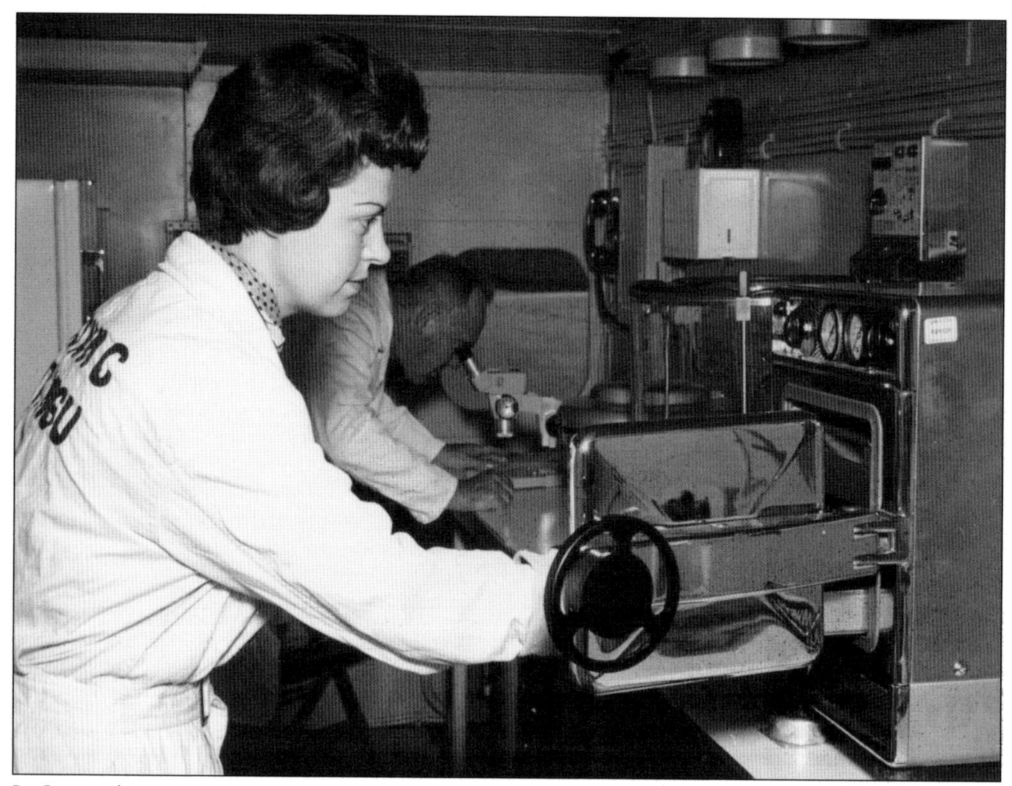

In September 1961, an unidentified biomedical laboratory technician places surgical instruments in a sterilizer while an unidentified space medic conducts an experiment with a dissectoscope, used when dissecting microscopic biological specimens. The van and equipment were part of a modular laboratory used to support missile and space programs on the Pacific Missile Range, which required life science support. (Courtesy of the National Archives.)

The biomedical van seen here in September 1961 was a prototype mobile bioscience laboratory designed by the PMR Bio-Medical Office to support biological equipment for use during missile and space programs conducted on the Pacific Missile Range. Composed of modular sections, which could be combined to provide different types of biomedical support, the vans were fabricated by the PMR Technical Support Directorate. (Courtesy of the National Archives.)

Measuring 85 feet in diameter, the massive TAA-2 antenna at Point Mugu was installed for use to relay via Syncom III satellite television coverage of the XVIII Olympic Games from Tokyo in 1964. The antenna was then reconfigured to track experimental satellites, including those for NASA's space program. In May 1972, the TAA-2 antenna was moved to New Mexico State University. (Courtesy of the National Archives.)

Project engineer James Pierce (left) and satellite geophysics officer Mark Jones (right) examine plastic foam material used to build a Geofoam hut (behind them) in 1968. Pierce and Jones developed the lightweight shelter to house scientific personnel and their instruments in remote regions of the world where climates are hostile and the trucking of conventional building materials was impossible. (Courtesy of the National Archives.)

The Mercury Station 13 team, located at Naval Missile Facility Point Arguello, tracks astronaut Lt. Col. John Glenn on his first pass while orbiting Earth on February 20, 1962. Members of Glenn's 1962 Mercury orbiting support team are, from left to right, Ted White, capsule systems observer; Buzz Aldrich, assistant capsule communicator; Wally Schirra, astronaut; Dr. Harry Bratt, and Dr. Carl Pruett, both biomedical observers. (Courtesy of the National Archives.)

The Targets Department located in the East Hangar Bay, Building 35, readies the target drone aircraft for flight on May 29, 1968. Point Mugu utilizes a large number of aircraft, missiles, and surface targets for use on the range as well as for testing and evaluating new equipment. (Courtesy of the National Archives.)

Radar tracking facilities located on Tranquillion Peak, NMF Point Arguello, are seen in 1962. The FPS-16 radar atop Tranquillion Peak overlooks Vandenberg Space Launch Complex 6 and the shoreline. The radar provides data and range safety for missile firings. More than 1,800 missiles and space boosters have been launched from Vandenberg Air Force Base since December 1958. (Courtesy of the National Archives.)

A December 20, 1962, view shows Point Mugu and Laguna Peak, home of the Pacific Missile Range, established in 1958, with Point Mugu as headquarters. It was an ideal geographic location for assessing missiles and remote-controlled aircraft: the vast expanse of the Pacific Ocean offered a safe area for test operations, and Laguna Peak provided a convenient spot for communications and instrumentation equipment, as did the islands for instrumentation equipment on the sea range. The triad—Point Mugu, San Nicolas, and Point Arguello—was one of the most heavily instrumented areas in the world at the time. Much of the fieldwork was conducted within this tightly controlled sea range. Air squadrons regularly fired missiles, and Marine Corps antiaircraft battalions launched missile batteries at target drones. (Courtesy of the US Navy Seabee Museum.)

Pictured from left to right, marine biologists Prof. George MacGinitie, his wife, Nettie, and Hospitalman 3rd Class James D. Byrley pause their scientific explorations at Mugu Lagoon at a sign warning against unauthorized visits. For decades, the Navy has protected Mugu Lagoon against trespassing with fences, guards, and signs. As a result, it has become a last refuge on the West Coast for rare biological specimens. (Courtesy of the National Archives.)

A June 30, 1959, photograph looking towards the south shows the San Nicolas Island runway shortly after expansion was completed. After years of using early-range communications and data systems, new modern transmitters and receivers, microwave systems, telemetry facilities, and radars were installed simultaneously with the runway expansion as part of a major upgrading project at both San Nicolas and Santa Cruz Islands. (Courtesy of the US Navy Seabee Museum.)

FPS-16 radars sit on Mount Tranquillon, overlooking all of Naval Missile Facility Point Arguello and Vandenberg Air Force Base around 1962. The radar provides data and range safety for missile launches. The FPS-16 radar at Vandenberg AFB, California, has been used for tracking NASA space vehicles since the 1960s. (Courtesy of the US Navy Seabee Museum.)

Seen in this c. 1966 photograph, the Naval Missile Center target trailer, located at Vandenberg Air Force Base, was used as part of the Bomarc target launch program. In 1964, the Navy accepted 126 surplus Bomarc A missiles from the Air Force, and preparations began to convert the missiles into targets for test research. The launch site was located at Vandenberg AFB because it was more accessible than the island launch complexes. (Courtesy of the National Archives.)

The first Nike-Zeus missile launch took place at Point Mugu in September 1961. Additional launches took place in rapid succession thereafter. The 450,000-pound thrust, solid propellant was the most powerful ever fired on the range to that point. After the 19th successful Nike-Zeus firing on the range in December 1962, the project was moved to the Pacific Missile Range Facility in Kwajalein, Marshall Islands, for the advance testing stage. (Courtesy of the National Archives.)

The commanding officer of Guided Missile Unit 41 (GMU-41) speaks at the change of command ceremony at Naval Air Station Point Mugu around 1970. Guided Missile Unit 41 provided support to VX-4 in the operational and environmental testing of air-to-air guided missiles and conventional ordnance. The unit provided technical support and telemetry data on missiles during training exercises. (Courtesy of the National Archives.)

A February 16, 1967, view shows Buildings 850–858 with FPS-16 radars mounted on them located along Beach Road past the end of the runway and ditch roads. Starting in 1958, the first FPS-16 radar specifically designed for instrumentation was installed at Point Mugu. Four systems were located at both Point Mugu and San Nicolas Island. The FPS-16 was periodically updated with improved modifications. (Courtesy of the National Archives.)

Naval Missile Center Point Mugu air operations are seen in 1966. As part of its support, NMC provided aerial and surface targets for the missile procedures conducted at Point Mugu and for ships engaged in training exercises. Flight evaluation also required many types of aircraft with various functions. Service crews performed maintenance with special training in servicing, repair, and maintaining electronic gear. (Courtesy of the National Archives.)

This April 14, 1966, image shows Laguna Peak with NAS Point Mugu in the background. In March 1963, the Navy installed a dish antenna atop the peak as part of the first space communications station designed to inject information into the Navy's first navigational satellite. The 1,200-pound reflector was 60 feet in diameter and was designed and constructed by Philco's Western Development Laboratories of Palo Alto, California. (Courtesy of the National Archives.)

A Thor-Able Star rocket at Cape Canaveral prepares to launch the Transit satellite, on April 14, 1960. The Navy Astronautics Group was commissioned in April 1962 to operate the Navy Navigational Satellite System, known as Transit. The satellite continuously broadcast satellite position messages, providing accurate, all-weather navigation capability to fleet ballistic missile submarines, naval forces, and civilian users. (Courtesy of the National Archives.)

A medium-gain telemetry antenna installed on San Nicolas Island to receive telemetry data from missiles and rockets fired on the sea test range is seen in 1968. San Nicolas Island became an essential link in the chain, although other offshore islands were used and developed in various ways. (Courtesy of Naval Base Ventura County.)

The Range Instrumentation Facility Timing Center control panel and readout located in Building 53 at Naval Missile Center Point Mugu is seen on February 12, 1968. In the late 1950s, maintaining precise time remained complicated, and the Pacific Missile Range (PMR) began synchronizing with the National Bureau of Standards by radio signal. By 1964, PMR set the timing center with the US Naval Observation Laboratory using the atomic standard. (Courtesy of the National Archives.)

Center personnel recover a Hydra II test vehicle from the Pacific Ocean in 1960. Project Hydra, the Navy's concept for launching large, solid-propellant rockets from the surface of the ocean, was born at Point Mugu. The Hydra concept involved floating the rocket vehicle vertically, similar to a spar buoy, on the ocean's surface prior to launch. The vehicle lifted directly from an upright floating position in the water. (Courtesy of the National Archives.)

Pictured in August 1968, Lt. Comdr. Harvey C. Sample, with the Range Operations Department, watches an air-to-air missile launch on a television monitor in the Range Communication Division telecommunications center while Western Electric chief installer Paul Carr monitors the new equipment. In 1967, construction was completed of the Range Communications Building, which housed the Telecommunications Switching System and microwave system. (Courtesy of the National Archives.)

A February 9, 1966, image of Point Mugu Missile Park showcases, from left to right, the Regulus II, Polaris, and Regulus missiles. The Point Mugu Missile Park is located alongside Pacific Coast Highway just outside Naval Base Ventura County Point Mugu. Constructed in 1966, the park houses a wide variety of the missiles and airplanes that have been tested at Point Mugu since World War II. (Courtesy of the National Archives.)

The Navy's deep-diving bathyscaph *Trieste* goes to sea again on October 10, 1965, in search of the sunken nuclear submarine USS *Thresher.* Transported to the Naval Electronics Laboratory's facility in San Diego, the *Trieste* was modified extensively and used in a series of deep-submergence tests in the Pacific Ocean, often near Point Mugu, culminating in a dive to the bottom of the Mariana Trench in January 1960. (Courtesy of the National Archives.)

The set of the television show *Mission: Impossible* moved to Point Mugu to film an episode during its sixth season in 1971. The original series was filmed almost exclusively around Hollywood and the Los Angeles Basin, as were many other shows during that period. Pasadena and the Caltech campus were common locations. Another noted site was the Bradbury Building used in other films and series. (Courtesy of the National Archives.)

The San Nicolas Island LCM (Landing Craft, Mechanized) target barge or range ship with newly installed radar equipment is seen here on June 26, 1968. Missile range instrumentation ships were equipped with antennae and electronics to support the launching and tracking of missiles and rockets. Since many missile ranges launched across ocean areas for safety reasons, the ships extended the range of shore-based tracking facilities. (Courtesy of the National Archives.)

Pres. Richard M. Nixon and First Lady Pat Nixon wave to the crowd as they disembark from Air Force One on March 21, 1969. The Nixons stopped at Point Mugu to award the Presidential Unit Citation to the Seabees of Naval Mobile Construction Battalion 3. The Seabees were awarded the citation for assisting the Marines in the defense of Hue City during the Tet Offensive in Vietnam. (Courtesy of the National Archives.)

Marines attached to the Marine Aviation Detachment show off their running skills for the inspector general's team visiting Point Mugu on May 14, 1970. The Marine Aviation Detachment at Point Mugu performed liaison duties between the Marine Corps and the local commands, administrative services and control for Marines assigned to the base, and Marine technical support billets needed at Point Mugu. (Courtesy of the National Archives.)

Sgt. R.W. Ault (left) and Cpl. J.B. Bray (right), with Marine Aviation Detachment Point Mugu, perform preoperational calibration on the AN/TPQ-10 system in the field. The Marine Aviation Detachment Point Mugu provides project management, aviation support, technical expertise, and fleet support for assigned Marine Corps weapons systems and related devices throughout the weapons systems life cycle. (Courtesy of the National Archives.)

The Coast Guard rushes a civilian pilot to a station ambulance so he can be taken to St. John's Hospital in Oxnard. The pilot, who went down with his single-engine plane in the Pacific Ocean south of San Nicolas Island, was rescued from the plane's wing by a Coast Guard helicopter on October 4, 1974, just before the downed aircraft sank. A Coast Guard C-130 located him shortly before he ditched the aircraft. (Courtesy of the National Archives.)

In 1960, the Navy Relief Fundraising Campaign launched the first Space Fair, the predecessor to the Point Mugu Air Show. The Space Fair featured the Blue Angels and exhibits by many of the country's leading missile and aircraft manufacturers, displays of Navy missiles and jet aircraft, and a carnival midway. (Courtesy of the US Navy Seabee Museum.)

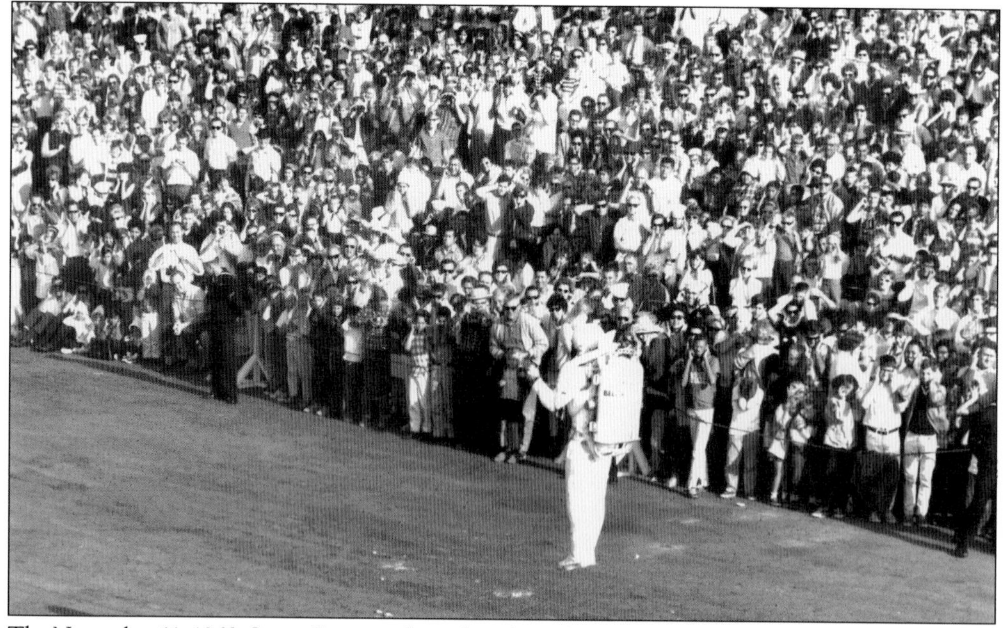

The November 11, 1968, Space Fair crowd watches the Bell Aerosystems Rocketbelt man demonstrate the newly developed jet pack at Point Mugu. The Bell Aerosystems Rocketbelt man flew twice a day November 9–11, 1968, at the Space Fair, which was the largest air show in the United States at the time. (Courtesy of the US Navy Seabee Museum.)

Much of base turned into a parking lot for the Naval Air Station Point Mugu Space Fair, held November 9, 1968 and attended by more than 195,000 people. Naval personnel were on hand to quickly guide 50,000 cars into the many free parking areas on the base. (Courtesy of the US Navy Seabee Museum.)

The carnival midway at the 1964 Space Fair was a quarter-mile of rides, games, and refreshments. Two huge hangars held aerospace and military exhibits. The grandstand accommodated 25,000 people for air shows during the Space Fair at Naval Air Station Point Mugu. More than 150,000 people attend the event annually. Created in 1960, the Space Fair evolved into the Point Mugu Air Show. (Courtesy of the US Navy Seabee Museum.)

The early hours begin to tell as fueling crews, finished with their work for a moment, grab a few winks on the catwalks along the USS *Ranger*'s flight deck. VX-4, known as "the Evaluators," frequently boarded aircraft carriers to test the suitability of an aircraft and its weapons. The unit developed tactics techniques and procedures to create Navy tactical manuals. (Courtesy of the National Archives.)

Naval Missile Center Technicians Bob Barnett (left) and Feliz E. Chavez (right) prepare the first production model of the Firebee II (BQM-34E) for wing load evaluation. The Threat Simulation Department evaluated the operational characteristics of the target before it was accepted for fleet-wide use. The Firebee II was a sleek dart aircraft with a swept tailplane and swept mid-body wings. (Courtesy of the National Archives.)

Naval personnel wash an A/3 aircraft with a new ND-390 mobile truck aircraft washer on May 8, 1968. Aircraft exterior maintenance included technical cleaning of landing gears and wheels and washing the total surface of the plane. Cleaning the aircraft and targets was part of the regular maintenance program in order to remove dirt, grease, and grime. (Courtesy of the National Archives.)

President Kennedy flashes his famous smile as he bids goodbye to the 60,000 people who turned out to see him during his brief stopover at Point Mugu on June 7, 1963. During his 30-minute stay at Naval Air Station Point Mugu, Kennedy addressed the crowd for a few minutes and then proceeded on his customary round of handshakes. (Courtesy of the National Archives.)

Eric I. Wasserman (left) and Duane L. Dodge, members of the Electromagnetic Technology Branch, mount a scale model of a T-33 aircraft for a radar reflectivity test in the anechoic chamber on March 18, 1968. An anechoic chamber is a room lined with microwave absorbing material in order to record data and measure scale models or weapons response to the elements. (Courtesy of the National Archives.)

Naval personnel and civilians work on an E-2A Hawkeye inside a Point Mugu hangar on July 7, 1964. The E-2A Hawkeye was designed to patrol the approaches around the fleet by detecting any impending attack by hostile aircraft, missiles, or sea forces. In addition to this function, the E-2A provided strike and traffic control, area surveillance, search and rescue guidance, navigational assistance, and communications relay services. (Courtesy of the US Navy.)

Four

A NEW GENERATION
1970–1991

Parked outside the Point Mugu hangars on August 22, 1974, four F-4s from the 49th Fighter Wing at Holloman Air Force Base, New Mexico, display their tail codes. The Pacific Missile Range provided range support to test the Air Force Systems Command's Airborne Warning and Control System (AWACS) program. AWACS was developed to provide an airborne surveillance capability in a modified Boeing 707 airframe. (Courtesy of the National Archives.)

An April 2, 1981, aerial view includes the Naval Air Reserve Unit (NARU) and Patrol Squadron 65 (VP-65) Building with Attack Squadron 305 (VA-305) and Light Attack Helicopter Squadron 5 (HAL-5) reserve units near the runways at the Pacific Missile Test Center. The units, which joined Point Mugu as tenant activities in January 1971, initially included 14 Skyhawks assigned to VA-305 and 12 Neptunes assigned to VP-65. (Courtesy of the US Navy.)

Navy Astronautics Group Explorer Post 2983 visits the Point Mugu airfield in 1975. Adults and youth members of the Explorer's group enjoy the static display that was part of the open house. The open house consisted of demonstrations showing the array of engineering and missile programs. The day's festivities peaked with an afternoon of aerial demonstrations. (Courtesy of the National Archives.)

Explorers visit the Point Mugu air traffic control tower in 1975. Point Mugu often hosted Boy Scout troops, Explorers, and open houses for youth groups and Navy dependent groups. These included a summer youth program for disadvantaged children started by Rear Adm. Howard Moore. A total of 250 boys and girls came to Point Mugu and CBC Port Hueneme for two weeks of tours, films, lectures, and sports. (Courtesy of the National Archives.)

A Harpoon missile launches from the Pad B Anti-Submarine Rocket (ASROC) launcher on December 1972. The photograph was shot by a remote camera set up by civilian George Addelman. The ASROC was developed by the Navy in the 1950s, deployed in the 1960s, and eventually installed on more than 200 Navy surface ships, specifically cruisers, destroyers, and frigates. (Courtesy of the National Archives.)

A Douglas EA-3B Skywarrior is parked at the Point Mugu airfield around 1968. The Skywarrior was a strategic bomber built for the Navy and is amongst the longest-serving carrier-based jet aircraft. It entered service in 1956 and was retired in 1991. For many years after its introduction, it was the heaviest aircraft ever flown from an aircraft carrier, earning it the unofficial nickname "the Whale." (Courtesy of the National Archives.)

Dr. S.L. Huang, right, and an unidentified assistant inspect a graphite epoxy composite after it was test-flown on the Point Mugu sea test range on September 6, 1973. A BQM-34E wing was fitted by the Naval Missile Center Thrust Simulation Department and flight-tested by Pacific Missile Range personnel. The graphite epoxy was lighter and the wing was stronger than its metal counterpart. It was also more resistant to corrosion and fatigue. (Courtesy of the National Archives.)

This 1970 view shows hangar 553 with QF-9 and QT-53 unmanned, remote-controlled drone aircraft parked on the targets' flight line, waiting to help train antiaircraft crews. Simplistically, target drones resemble radio-controlled model aircraft in missile shaped exteriors. Modern drones often use counter measures, radars, and similar devices to mimic real aircraft. (Courtesy of the US Navy.)

A row of LTV A-7 Corsair II's is parked on the flight line at NAS Point Mugu on March 1, 1976. The aircraft are painted in a bicentennial theme, commemorating America's 200th birthday. (Courtesy of the US Navy.)

Comdr. Rosemary Mariner assumes command of Tactical Electronic Warfare Squadron (VAQ-34) at Naval Air Station Point Mugu, California, in 1990. Mariner became the first Navy woman to assume command of an aviation squadron. In 1993, she became one of the first female aviators to be promoted to captain. (Courtesy of the US Navy.)

This October 1, 1971, view of the parking apron and hangars from the tower looks southeast. At Point Mugu, missiles were mated with new aircraft and weapon systems were tried out on the range. The Lark, Loon, Gargoyle, and many more missiles were tested at the center, and each weapon accelerated missile research and development, testing, and evaluation programs. (Courtesy of the National Archives.)

Congressman Robert W. Daniel Jr., a member of the House Armed Services Committee, arrives at Point Mugu on November 11, 1974, for an F-14 orientation flight. Daniel flew into Mugu in a VF-51 F-4 Phantom from Naval Air Station Miramar. Lt. Comdr. Ron Ludlow, VX-4, flew Daniel on a one-and-one-half hour flight over the Pacific Missile Range Sea Test Range. (Courtesy of the National Archives.)

Seen here on August 21, 1981, tracking and antenna stations are located on Santa Cruz Island, just off the coast of California near the Pacific Missile Test Center. The Navy utilizes a mountaintop near the eastern end of the island for an instrumentation complex, housed on a 10-acre parcel and including barracks, a power plant, fire station, and heliport. (Courtesy of the US Navy.)

Public Works Maintenance and Transportation Division personnel demolish the original ACORN Assembly and Training Detachment hangar in July 1975. It served as the Naval Air Missile Test Center's only hangar until the construction of Hangar 34 in 1949. In May 1956, it was converted into a fire station until August 1971 and was later used by VXE-6 as a helicopter hangar. (Courtesy of the National Archives.)

Naval Air Station Point Mugu Fire Department Station 72, with new engines, opened around 1970. Seventy-nine firefighters are assigned to Naval Base Ventura County, and each must have expertise in structure, airfield, and shipboard fires, as well as in local issues including wildfires and emergency medical care. The firefighters also teach lifesaving skills to school children and provide the local community with training and safety classes. (Courtesy of the National Archives.)

In this December 29, 1971, image, the antenna array of the USS *Wheeling* shows the AN/FPS-16 radar, the AN/FPS-25 radar, the command destruct, the low-gain telemetry antenna, the medium-gain telemetry antenna, and the log periodic antenna. The *Wheeling* was converted into a range instrumentation ship in 1962 and assigned to the Military Sea Transportation Service to support operations on the Pacific Missile Range on May 28, 1964. (Courtesy of the National Archives.)

Girls Scouts pose near one of the painted bicentennial fire hydrants on March 8, 1976. The Girls Scouts held a contest to paint the base's fire hydrants with a bicentennial theme. The winners of the contest are, from left to right, Stacey Deckart, Troop 8; Karen Clasen, Troop 150; Deborah Beason, Troop 590; Bernadette Steelman, Troop 150; and Angela Richards, Troop 8. (Courtesy of the National Archives.)

Pres. Ronald Reagan chats with the base commander after landing at the airfield in the 1980s. The president and first lady used Naval Air Station Point Mugu quite frequently during Reagan's tenure due to its proximity to the Reagan Ranch in Santa Barbara. During his presidency, Reagan spent vacations at the ranch, which became known as "the Western White House." (Courtesy of the US Navy.)

Adm. James L. Holloway III visits Point Mugu in 1974. As Chief of Naval Operations from 1974 to 1978, Holloway was a member of the Joint Chiefs of Staff (JCS) and served as chairman of the JCS during the evacuation of Cyprus, the rescue of the merchant ship SS *Mayaguez* and its crew, and the punitive strike operations against the Cambodian forces involved in the *Mayaguez*'s seizure. (Courtesy of the National Archives.)

Actors Edmund Burr (left) and John Rayburn (partially visible at right) watch a Japanese plane attack the air tower during filming of the Universal Pictures movie *Midway*, on July 10, 1975. The main filming at Point Mugu centered on the attack and eventual destruction of the wooden air control tower of Midway's wartime airstrip. (Courtesy of the National Archives.)

An LC-130 Hercules aircraft with the Navy's Antarctica Development Squadron 6 (VXE-6) taxis past an Air Force C-141B Starlifter aircraft on November 16, 1988. VXE-6 was established on January 17, 1955, at Naval Air Station Patuxent River, Maryland, as part of the Navy's evolving role of providing support for scientific exploration of Antarctica. The squadron's home base changed to NAS Point Mugu in 1974. (Courtesy of the US Navy.)

In a June 15, 1988, photograph, members of VXE-6 prepare to offload pallets of supplies from an LC-130 Hercules aircraft at a remote research camp during Operation Deep Freeze. During its service, VXE-6 transported more than 195,000 passengers, 240 million pounds of dry cargo, and almost 10 million gallons of fuel to sites in Antarctica. (Courtesy of the US Navy.)

This view is of the helicopter landing zone and hangar at Palmer Station, Antarctica, with four UH-1N Iroquois helicopters from the Navy's Antarctic Development Squadron 6 (VXE-6), November 16, 1988. Since its establishment, VXE-6 logged more than 200,000 flight hours in direct support of the United States' interests in Antarctica. (Courtesy of the US Navy.)

An April 21, 1970, aerial view shows radar lined up along Makaha Ridge near Barking Sands, Kauai, Hawaii. The Pacific Missile Range Facility Barking Sands (PMRF) is the world's largest instrumented multi-environment range capable of supporting surface, subsurface, air, and space operations simultaneously. The facility consists of more than 1,100 square miles of instrumented underwater range and more than 42,000 square miles of controlled airspace. (Courtesy of the National Archives.)

A break in the levee at Calleguas Creek on February 18, 1980, caused water to come straight onto Naval Air Station Point Mugu. After it broke, the Calleguas Creek channel, which was much higher that the surrounding farmland, flooded the farmlands along the stream. Floodwater then reentered the channel through yet another break downstream after flooding the base. (Courtesy of the US Navy Seabee Museum.)

On February 19, 1980, Navy dependents collected and arranged donations for flood victims. The full shock of the flood hit hardest for the evacuees after they returned home to discover destroyed cars, carpets, furniture, clothing, and personal heirlooms. The personal property damage suffered by families was approximately $9 million. (Courtesy of the US Navy Seabee Museum.)

Flood damage in the Naval Air Station Point Mugu housing area is seen here on February 19, 1980, after a Calleguas Creek levee broke. The damaged levee unleashed its devastating stock of water and resulted in the flooding of more than 80 percent of Naval Air Station Point Mugu on February 18, 1980. None of the 550 homes at Point Mugu escaped damage. (Courtesy of the US Navy Seabee Museum.)

Point Mugu naval personnel and dependents assemble at the Naval Construction Battalion Center Port Hueneme Housing Office for new home assignments on February 19, 1980. Buses and trucks began transporting 3,000 evacuees to temporary quarters at Port Hueneme within hours of the levee collapsing. A group of NMCB-5 Seabees assisted evacuees in moving salvaged personal articles to temporary quarters. (Courtesy of the US Navy Seabee Museum.)

This Pacific Missile Range (PMR) float participated in the 13th Annual Armed Forces Day Parade and celebration in Torrance, California, on January 27, 1972. Throughout the week of the parade, the PMR photograph exhibit outlining the functions of the various commands at Point Mugu was on display in the Del Amo Fashion Square in Torrance. The display was manned by Photographers Mate 3rd Class Donald Odom and Aviation Ordnanceman 2nd Class W.F. Freitas. (Courtesy of the National Archives.)

The main gate sign is seen in 1975, after the base name changed to the Pacific Missile Test Center. On April 25, 1975, the Naval Missile Center, Pacific Missile Range, and Naval Air Station Point Mugu merged into the Pacific Missile Test Center. The merger occurred to consolidate base operations and range, test, and evaluation activities, and it helped personnel to focus on a single purpose. (Courtesy of the US Navy.)

Five

Transformation 1992–2010

The famous Laguna Peak telemetry antenna and radar equipment is lit up at night. The Naval Satellite Operations Center (NAVSOC) recently completed a Service Life Enhancement Project for the 60-foot antenna on Laguna Peak. This $1-million project replaced the 38-year-old hydraulic pointing system with a modern electromechanical pointing system. This valuable antenna now supports NAVSOC operations. (Courtesy of the National Archives.)

The front gate changed its sign after the Naval Air Warfare Center was commissioned in 1992. On January 22, 1992, the Pacific Missile Test Center was disestablished, and the Naval Air Warfare Center was formed. This aligned technical functions with those of Naval Air Weapons Center China Lake. (Courtesy of the US Navy.)

Unidentified sailors from Naval Base Ventura County Point Mugu stand on the base's missile float and wave to spectators during the Fourth Annual San Fernando Valley Veterans Day Parade, on November 11, 2007. The parade tradition started in 2004 when Congressman Berman, Councilman Padilla, and Commissioner Flores worked with members of the community to establish the event. (Courtesy of the US Navy.)

A crowd of nearly 74,000 people watched as the Air Force's premier precision flying team, the Thunderbirds, push their F-16 Fighting Falcon aircraft to the limit while performing at the Naval Base Ventura County Air Show at Point Mugu on April 1, 2007. The squadron tours the United States and much of the world, performing aerobatic formation and solo flying in specially-marked Air Force jet aircraft. (Courtesy of the US Navy.)

On August 15, 2008, riders assigned to Naval Base Ventura County's Midsummer Motorcycle Ride for Safety program return to Point Mugu after traveling in three groups to the Deer Lodge in Ojai to promote motorcycle safety awareness and experience rider mentorship. Approximately 60 military personnel and Department of Defense employees participated in the ride. (Courtesy of the US Navy.)

October 11, 2000, marked the establishment of Naval Base Ventura County (NBVC) during a ceremony held at Point Mugu. The two commands of Naval Air Station Point Mugu and Construction Battalion Center Port Hueneme were consolidated into one organization that is one of the major naval installations on the West Coast. (Courtesy of the US Navy.)

An unidentified aircraft flight deck director guides an E-2C Hawkeye assigned to the "Golden Hawks" of Carrier Airborne Early Warning Squadron One One Two (VAW-112) out of the landing area after making an arrested gear landing on the flight deck of the USS *John C. Stennis* (CVN-74) on January 22, 2007. The *Stennis* conducted carrier qualifications prior to heading west to help bolster security in the US Central Command area of operations. (Courtesy of the US Navy.)

An E-2C Hawkeye assigned to the Golden Hawks is parked in front of the Island aboard the Nimitz-class aircraft carrier USS *John C. Stennis* in preparation for launch during nighttime flight operations on January 28, 2007. (Courtesy of the US Navy.)

In this image taken on November 20, 2003, Aviation Structural Mechanic 3rd Class Seth Ignasiak, while aboard the USS *John C. Stennis*, displays "the Judge," an oversized sledgehammer used for positioning lifting jacks under aircraft. Ignasiak, the E-2C Hawkeye maintenance crew, and the crew of the Black Eagles of VAW-113, were conducted Composite Unit Training Exercises (COMPTUEX) aboard the *Stennis* in preparation for its upcoming deployment. (Courtesy of the US Navy.)

An aircraft flight deck director guides an E-2C Hawkeye assigned to the Golden Hawks onto a steam-propelled catapult for launch off the flight deck of the Nimitz-class aircraft carrier USS *John C. Stennis* on January 22, 2007. (Courtesy of the US Navy.)

On March 14, 2009, Seabees assigned to Naval Mobile Construction Battalion 5 (NMCB-5) board a commercial aircraft at Point Mugu for transportation to Kuwait, where they will receive their final gear issue before departing for Afghanistan. (Courtesy of the US Navy.)

Aviation Structural Mechanic Airman Russell Miller greases the landing gear of an E-2C Hawkeye, assigned to the Black Eagles of Carrier Airborne Early Warning Squadron One One Three (VAW-113), on the flight deck of the aircraft carrier USS *Ronald Reagan* (CVN-76) on October 29, 2005. The *Ronald Reagan* was underway in the Pacific Ocean participating in the ship's initial Composite Unit Training Exercise (COMPTUEX). (Courtesy of the US Navy.)

An unidentified sailor performs maintenance to an E-2C Hawkeye assigned to "the Sun Kings" of Carrier Airborne Early Warning Squadron One One Six (VAW-116) aboard the Nimitz-class aircraft carrier USS *Abraham Lincoln* (CVN-72) on April 3, 2006. The *Lincoln* and Carrier Air Wing Two (CVW-2) were underway in the western Pacific area of operations. (Courtesy of the US Navy.)

Aviation Structural Mechanic Airman Elliot Littles cleans the underside of an E-2C Hawkeye assigned to VAW-116 in the ship's hangar bay aboard the aircraft carrier USS *Abraham Lincoln* on March 31, 2006. The *Lincoln* and Carrier Air Wing Two (CVW-2) were underway in support of exercises Foal Eagle, Valiant Shield, and Rimpac. (Courtesy of the US Navy.)

Aircrew Survival Equipmentman 3rd Class Vincente Perez (left) and Aircrew Survival Equipmentman 1st Class Gregory Majors service a parachute from an E-2C Hawkeye assigned to the Black Eagles of VAW-113, in the paraloft aboard the USS *John C. Stennis* on June 17, 2004. (Courtesy of the US Navy.)

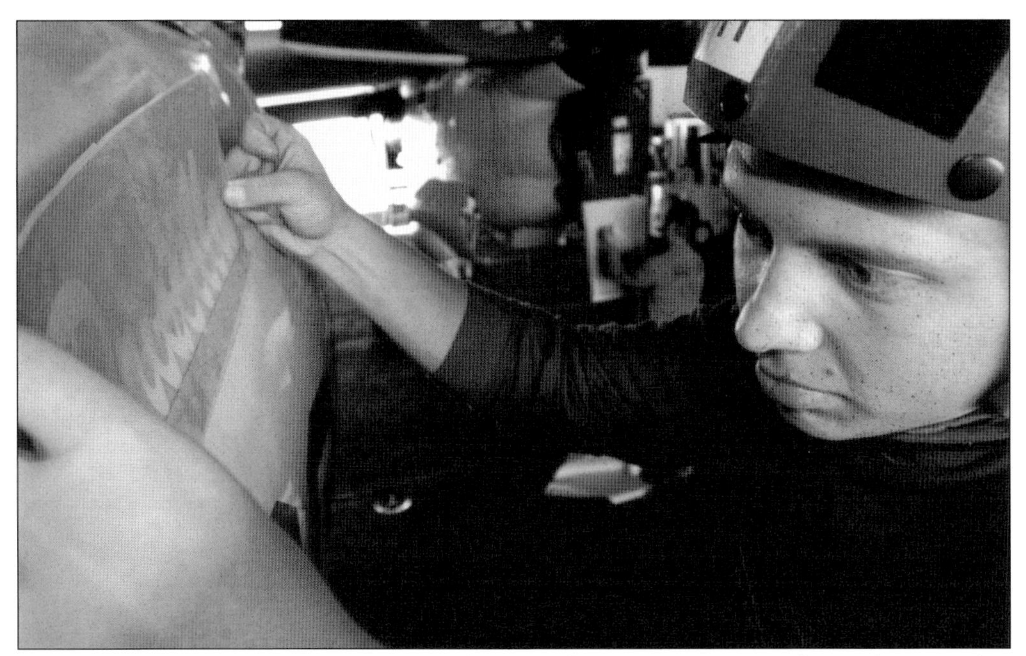

Aviation Structural Mechanic Airman Will Morse places a painting stencil onto the nose of an E-2C Hawkeye assigned to the Sun Kings of VAW-116 in the hangar bay aboard the USS *Abraham Lincoln* on February 16, 2005. On April 20, 1967, VAW-116 was commissioned to pilot and maintain the E-2B aircraft while a part of Carrier Air Wing 15. (Courtesy of the US Navy.)

An E-2C Hawkeye assigned to the Sun Kings places engines in full military power aboard the Nimitz-class aircraft carrier USS *Abraham Lincoln* prior to flight operations on December 7, 2004. The carrier was operating in Southern California waters to enable aircraft to accomplish as much accelerated training as possible before combat missions. (Courtesy of the US Navy.)

On June 15, 2005, an unidentified pilot assigned to the Sun Kings of VAW-116 conducts preflight checks on one of the squadron's E-2C Hawkeyes prior to flight operations aboard the USS *Abraham Lincoln*. The *Lincoln* was at sea conducting readiness training in support of the Navy's Fleet Response Plan (FRP). (Courtesy of the US Navy.)

In Banda Aceh, Sumatra, Indonesia, Aviation Structural Mechanic Airman Vincent Estrada, assigned to VAW-116, helps load bags of rice onto a Navy helicopter on January 16, 2004. The humanitarian aid supplies were being delivered to the victims of the Southeast Asia tsunami as part of Operation Unified Assistance. (Courtesy of the US Navy.)

Aviation Machinist's Mate 1st Class Peter Farala (left) and Aviation Machinist Mate 3rd Class Tran Quy, assigned to the Sun Kings, install a propeller on an E-2C Hawkeye after performing scheduled maintenance on September 6, 2004. Embarked on the USS *Abraham Lincoln*, VAW-116 conducted operations prior to deployment. (Courtesy of the US Navy.)

On March 22, 2005, Aviation Boatswain's Mate 3rd Class Jeffery Jaca prepares an E-2C Hawkeye, assigned to "the Wallbangers" of Carrier Airborne Early Warning Squadron One One Seven (VAW-117) for launch on the flight deck of the USS *Nimitz* (CVN-68). The *Nimitz* and Carrier Strike Group Eleven (CSG-11) were conducting a Joint Task Force Training Exercise (JTFEX) off the coast of Southern California. (Courtesy of the US Navy.)

Lt. Nathaniel Dishman, an arresting gear officer, stands watch to ensure the flight deck is clear for landing as an E-2C Hawkeye assigned to the Wallbangers of VAW-117 lands aboard the USS *John C. Stennis* on July 13, 2006. The *Stennis* was underway conducting carrier qualifications off the coast of Southern California. (Courtesy of the US Navy.)

Personnel assigned to the Wallbangers install a propeller pump housing on an E-2C Hawkeye engine in the hangar bay aboard the aircraft carrier USS *Nimitz* on October 27, 2005. Approximately 150 officers and enlisted personnel make up the Wallbangers team. While their backgrounds vary, together they form a powerful fighting squadron. (Courtesy of the US Navy.)

Airman Dustin Roach washes down the wing of an E-2C Hawkeye assigned to the Wallbangers of VAW-117 aboard the USS *Ronald Reagan*, on June 30, 2004. The *Reagan* was returning home from participating in exercises supporting Summer Pulse 2004, which was a demonstration of the Navy's ability to provide credible combat across the globe with allied forces. (Courtesy of the US Navy.)

An E-2C Hawkeye assigned to the Sun Kings of VAW-116 is hoisted from the USS *Constellation* (CV-64) and lowered onto a barge in Bahrain on February 20, 2003. The *Constellation* was deployed to the Persian Gulf in support of Operation Enduring Freedom. (Courtesy of the US Navy.)

An E-2C Hawkeye assigned to the Wallbangers receives a wash down from Aviation Electrician's Mate Dawson Lorton while aboard the USS *Ronald Reagan* on June 7, 2004. The USS *Ronald Reagan* was in the south Atlantic Ocean circumnavigating South America during its transit to a new home port in San Diego. (Courtesy of the US Navy.)

On October 8, 2008, Senior Chief Aviation Mechanic Jan Hirschfeld, assigned to the Sun Kings of VAW-116, is greeted by his children after his return from a deployment aboard the USS *Abraham Lincoln*. The USS *Abraham Lincoln* was deployed to the Fifth Fleet area of responsibility supporting maritime security operations. (Courtesy of the US Navy.)

Seabees with the 31st Seabee Readiness Group load part of a Scalable, Modular, Agile, Responsive Table of Allowance aboard the back of a US Air Force C-5 Galaxy at Naval Base Ventura County Point Mugu on January 27, 2010. The Table of Allowance (TOA) outfits the naval mobile construction battalions with the capability to perform building operations under contingency conditions for 90 days without being resupplied. (Courtesy of the US Navy.)

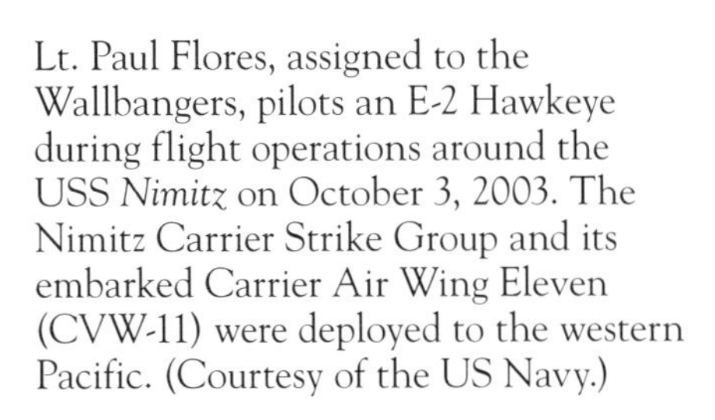
Lt. Paul Flores, assigned to the Wallbangers, pilots an E-2 Hawkeye during flight operations around the USS *Nimitz* on October 3, 2003. The Nimitz Carrier Strike Group and its embarked Carrier Air Wing Eleven (CVW-11) were deployed to the western Pacific. (Courtesy of the US Navy.)

On the flight deck aboard the nuclear-powered aircraft carrier USS *Nimitz* on August 5, 2005, Airman Andrew Clarke, assigned to the Wallbangers, signals a pilot to start the engines on an E-2C Hawkeye. The Nimitz Strike Group was on a regularly scheduled deployment to the Persian Gulf participating in maritime security operations. (Courtesy of the US Navy.)

Master Chief Petty Officer of the Navy Rick West (right) meets sailors during his tour of the Fleet Readiness Center while on a visit to Naval Base Ventura County on December 9, 2009. West became Master Chief Petty Officer of the Navy on December 12, 2008, after serving as Fleet Master Chief for the Pacific Fleet. (Courtesy of the US Navy.)

An E-2C Hawkeye assigned to the Wallbangers launches from one of four steam-powered catapults on the flight deck aboard the aircraft carrier USS *Nimitz* on December 2, 2004. The USS *Nimitz* was conducting Composite Training Unit Exercises off the coast of Southern California. (Courtesy of the US Navy.)

Karen Barton, part of the Morale, Welfare and Recreation Office at Naval Base Ventura County, demonstrates one of the fitness stations to personnel on July 24, 2008. Barton runs the Navy Fitness Van, an outreach program that assists sailors with staying in shape. Morale, Welfare and Recreation also offers aerobic classes, fitness centers, swimming lessons, aquatic activities, bowling, athletic fields and courts, youth activities, and golf. (Courtesy of the US Navy.)

The world's first open-ocean, roll-on roll-off pier, located on San Nicolas Island about 70 miles off the California coast, is under construction on September 21, 2004. The $12-million construction project on Naval Outlying Landing Field San Nicolas Island enabled the Navy to safely offload equipment and supplies for use by the island's Navy and civilian populations. (Courtesy of the US Navy.)

A US military multiservice ceremonial honor guard removes the flag-draped coffin of former president Ronald Reagan from a hearse at the Point Mugu airfield and prepares to place it aboard US Air Force VC-25 Special Airlift Mission 28000 on June 9, 2004. The president's body was flown to Washington, DC, where it lay in state in the Capitol rotunda, followed by a state funeral conducted at Washington National Cathedral. (Courtesy of the US Navy.)

An Arrow antiballistic missile interceptor is launched from its mobile platform during a joint Israeli-American developmental test at the Point Mugu Sea Range on August 26, 2004. This was the second in a series of appraisals following a previous successful test in which a target that represents a real threat to Israel was intercepted and destroyed. The test was part of the ongoing Arrow System Improvement Program. (Courtesy of the US Navy.)

A Tactical Tomahawk, the next generation of Tomahawk cruise missiles, is launched during a contractor test and evaluation on August 23, 2002. The Tomahawk missile provides a long-range, highly survivable, unmanned land attack weapon system capable of pinpoint accuracy. This new Tomahawk includes a launch-platform, mission-planning capability; in-flight retargeting; battle damage assessment capability; and in-flight health and status reporting through a satellite data link. (Courtesy of the US Navy.)

During a training exercise on May 5, 2004, an E-2C Hawkeye assigned to the Sun Kings of VAW-116 (top) flies in formation with an E-2C Hawkeye assigned to the Wallbangers of VAW-117 (bottom) off the coast of Southern California. (Courtesy of the US Navy.)

On April 5, 2004, Martin Ruane, a base ecologist onboard Naval Base Ventura County, conducts routine observations and surveys to monitor the different animal species that have made habitats in and around Mugu Lagoon at Naval Base Ventura County Point Mugu. The lagoon, which is the largest coastal wetland in Southern California, provides refuge for a variety of sensitive, wetland-dependent flora and fauna. (Courtesy of the US Navy.)

Machinery Technician 2nd Class Matthew Merel cuts on the lip of a wave during a preliminary heat in the military men's division of the Point Mugu Surf Contest at Naval Base Ventura County on August 22, 2009. Merel went on to place first in the finals. The Quiksilver-sponsored contest was open to both civilians and military personnel. (Courtesy of the US Navy.)

Mobile Utilities Support Equipment (MUSE) personnel load a power unit onto an Air Mobility Command C-5 Galaxy cargo plane around 2002. The Navy's MUSE program supports shore establishment utility systems and cold-iron services throughout the world with emergency power units, equipment capable of steam and electrical generation, and electrical conversion. (Courtesy of the US Navy.)

An E-2C Hawkeye assigned to VAW-113 makes a low pass during practice for an air-power demonstration aboard the Nimitz-class aircraft carrier USS *Ronald Reagan* on June 27, 2006. The *Ronald Reagan* and embarked Carrier Air Wing 14 (CVW-14) deployed to conduct maritime security operations as part of the global war on terrorism. (Courtesy of the US Navy.)

From the cockpit of his F-16 Fighting Falcon at Naval Base Ventura County Point Mugu, US Air Force Maj. Mark Arnholt, of the New Mexico Air National Guard (NMANG), reviews the preflight checklist prior to takeoff on July 15, 2007. The NMANG was one of several units participating in the defense system evaluation at Naval Base Ventura County. (Courtesy of the US Navy.)

Seabees from Naval Mobile Construction Battalion Forty (NMCB-40) load equipment onto an Air Mobility Command C-5 Galaxy cargo plane at the Point Mugu airfield on December 13, 2002. NMCB-40 deployed to Guam in support of disaster relief efforts after Super Typhoon Pongsona passed over the island on December 8, 2002. (Courtesy of the US Navy.)

The Firehawks crew from Helicopter Combat Support Squadron Five (HCS-5) looks over its 7.63-mm M134 Minigun machine gun mounted on a SH-60H Seahawk helicopter on May 1, 2001. The Firehawks were at the Shoalwater Bay training area to support special warfare operations during Exercise Tandem Thrust, a combined American, Australian, and Canadian military training exercise. (Courtesy of the US Navy.)

On September 16, 2009, Master-at-Arms 1st Class Blake Soller leads Jake, his military working dog, through training in narcotic detection onboard Naval Base Ventura County. The military working dog unit onboard Naval Base Ventura County was named Kennel of the Year for Commander, Naval Region Southwest. (Courtesy of the US Navy, photograph by Vance Vasquez.)

Troubleshooters from VAW-113 give the go ahead for takeoff from the flight deck aboard the USS *Abraham Lincoln* on March 24, 2003. In 2003, while on its way home, the *Lincoln* was turned around and sent back to the Persian Gulf. By mid-March, Operation Iraqi Freedom began, with the Black Eagles at the forefront of the conflict. (Courtesy of the US Navy.)

Members of Naval Mobile Construction Battalion Four (NMCB-4) load Tricon containers with construction equipment destined for field-testing in Iraq into a US Air Force Air Mobility Command C-5 Galaxy transport aircraft on August 31, 2006. The C-5 Galaxy, operated by the Air Force since 1969, provides strategic heavy airlift and carries oversized cargo across intercontinental distances. (Courtesy of the US Navy.)

Sailors from Naval Base Ventura County took part in the Law Enforcement Torch Run Special Olympics Southern California Relay on June 10, 2010. Sailors started the relay from Missile Park at Naval Base Ventura County Point Mugu and proceeded on a three-mile run to Mugu Rock. (Courtesy of the US Navy.)

A C-17 Loadmaster from March Air Force Base lands behind the Air Force Demonstration Team "the Thunderbirds" after showing the Loadmaster's capabilities to the crowd at the Naval Base Ventura County Air Show held at Point Mugu on August 8, 2010. A crowd estimated at more than 100,000 visited Naval Base Ventura County August 7–8, 2010, for the NBVC Air Show at Point Mugu. (Courtesy of the US Navy.)